ASCE Manuals and Reports on Engineering Practice No. 74

Guidelines for Electrical Transmission Line Structural Loading

Fourth Edition

Edited by Frank Agnew, P.E.

Task Committee on
Electrical Transmission Line Structural Loading

ASCE AMERICAN SOCIETY OF CIVIL ENGINEERS

SEI STRUCTURAL ENGINEERING INSTITUTE ASCE

Published by the American Society of Civil Engineers

Library of Congress Cataloging-in-Publication Data

Names: Agnew, Frank, editor. | American Society of Civil Engineers. Task Committee on Structural Loadings, author.
Title: Guidelines for electrical transmission line structural loading / Task Committee on Electrical Transmission Line Structural Loading, edited by Frank Agnew, P.E.
Description: Fourth edition. | Reston, Virginia : American Society of Civil Engineers, [2020] | Includes bibliographical references and index. | Summary: "MOP 74, Fourth Edition, provides up-to-date design and loading concepts, and applications specific to transmission line design"-- Provided by publisher.
Identifiers: LCCN 2020018035 | ISBN 9780784415566 (hardcover) | ISBN 9780784483084 (adobe pdf)
Subjects: LCSH: Electric lines--Poles and towers--Design and construction.Load factor design.
Classification: LCC TK3242 .G77 2020 | DDC 621.319/22--dc23
LC record available at https://lccn.loc.gov/2020018035

Published by American Society of Civil Engineers
1801 Alexander Bell Drive
Reston, Virginia 20191-4382
www.asce.org/bookstore | ascelibrary.org

Errata: Errata, if any, can be found at https://doi.org/10.1061/9780784415566.

ISBN 978-0-7844-1556-6 (print)
ISBN 978-0-7844-8308-4 (PDF)

Manufactured in the United States of America.

26 25 24 23 22 21 20 1 2 3 4 5

Photo credit: Ice photo on cover courtesy of AEP Transmission.

MANUALS AND REPORTS ON ENGINEERING PRACTICE

(As developed by the ASCE Technical Procedures Committee, July 1930, and revised March 1935, February 1962, and April 1982)

A manual or report in this series consists of an orderly presentation of facts on a particular subject, supplemented by an analysis of limitations and applications of these facts. It contains information useful to the average engineer in his or her everyday work, rather than findings that may be useful only occasionally or rarely. It is not in any sense a "standard," however, nor is it so elementary or so conclusive as to provide a "rule of thumb" for nonengineers.

Furthermore, material in this series, in distinction from a paper (which expresses only one person's observations or opinions), is the work of a committee or group selected to assemble and express information on a specific topic. As often as practicable the committee is under the direction of one or more of the Technical Divisions and Councils, and the product evolved has been subjected to review by the Executive Committee of the Division or Council. As a step in the process of this review, proposed manuscripts are often brought before the members of the Technical Divisions and Councils for comment, which may serve as the basis for improvement. When published, each manual shows the names of the committees by which it was compiled and indicates clearly the several processes through which it has passed in review, so that its merit may be definitely understood.

In February 1962 (and revised in April 1982), the Board of Direction voted to establish a series titled "Manuals and Reports on Engineering Practice" to include the manuals published and authorized to date, future Manuals of Professional Practice, and Reports on Engineering Practice. All such manual or report material of the Society would have been refereed in a manner approved by the Board Committee on Publications and would be bound, with applicable discussion, in books similar to past manuals. Numbering would be consecutive and would be a continuation of present manual numbers. In some cases of joint committee reports, bypassing of journal publications may be authorized.

A list of available Manuals of Practice can be found at http://www.asce.org/bookstore.

CONTENTS

PREFACE

The American Society of Civil Engineers Task Committee on Electrical Transmission Line Structural Loading provides design guidance to industry practitioners through the Manuals and Reports on Engineering Practices. This document, Manual of Practice No. 74, Fourth Edition, is intended to provide the most relevant and up-to-date information related to transmission line structural loading. It is not intended to be a step-by-step manual or a prescriptive code for direct implementation. Rather, it is intended to be a resource for development of a loading philosophy for electrical transmission structures which can be applied to an individual project or at a regional level. Much of the information contained within this document can be simplified for particular applications once regional or local climatic data and reliability levels are determined. The previous editions (1984, 1991, and 2010) have been well received and found wide use as practical guides to supplement mandatory legal state minimums. Although this Manual of Practice focuses on applications within the United States, the concepts presented are applicable worldwide.

In 2012, the ASCE Structural Engineering Institute Committee on Electrical Transmission Structures recognized the need for updates and revisions to Manual of Practice No. 74, Third Edition. The initial intent of the task committee was to update only sections of the manual affected by changes to national standards, particularly ASCE 7. As the task committee commenced review of the impacted sections, they recognized numerous sections within the manual for which present-day research and recent industry experience could be applied to significantly improve the content and organization of the manual. Thus, this resulting fourth edition was generally rewritten from the third edition.

There are several major concept changes in the updated Chapter 1 "Overview of Transmission Line Structural Loading" and Chapter 2 "Weather-Related Loads." The first of these is the decision to recommend a 100-year mean recurrence interval (MRI) as the basis for design and providing the corresponding wind speed and ice thickness maps for the United States. Additional wind speed maps and combined ice thickness and wind maps for 50-year and 300-year MRIs are provided in Appendix L. The additional maps have been included to allow users of this Manual of Practice to apply wind and ice loads associated with other MRIs as the previous method for translating loads between MRIs has been discontinued by ASCE 7. Chapter 2 includes some significant changes to components of the wind pressure formula, along with an extended discussion of high-intensity winds, such as downbursts and tornadoes, included in Appendix K.

Chapter 3 "Additional Load Considerations" and Chapter 4 "Wire System" have been enhanced with additional photos, graphs, and diagrams to give users of this Manual of Practice a better understanding of the loading concepts and application methodology as presented. Discussions have also been added to introduce additional loading cases as well as to elaborate on other important transmission structural loading concepts contained herein.

Chapter 5, "Examples," has been retained in this edition. The examples given have been updated to show the methodology of the changes within other chapters of the document. Chapter 5 has also been expanded in order to give the user additional guidance on key concepts presented elsewhere in this document.

Early in the task committee's work, there was a realization that the electrical transmission line industry would benefit from the development of a loading standard. As a result, an initial draft of a Transmission Line Structural Loading Standard document is included in Appendix M of this edition. This stand-alone draft Pre-Standard is included in this edition in order for transmission line owners, practitioners, and the public to comment on the content and form.

The recommendations presented herein reflect the consensus opinion of the task committee members and are applicable in the context of transmission line structural loading. Although intended as a guide for lines 69 kV and greater, the application of the concepts in this document might be justified at all voltages. The subject matter of this guide has been thoroughly researched; however, it should be applied only in the context of sound engineering judgment.

Committee Members

Roberto Behncke
Yair Berenstein
Adam Beyer
Ryan Bliss
Gary Bowles
Clinton Char
Ashraf El Damatty
Meihuan Zhu Fulk
Majid Farahani
Guy Faries
Joe Hallman
David Hancock
Robert Kluge
Paul Legrand
Ajay K. Mallik
Thomas G. Mara
Julie Matlage
James McGuire
Miguel Mendieta
Jacob Merriman
Michael Miller
Garett Muranaka
Dave Parrish
Jeremy Pettus
Scott Walton
C. Jerry Wong
Douglas Zylstra

Respectively submitted,
Task Committee on Structural Loadings
Frank Agnew, *Chair*
Ron Carrington, *Vice Chair*
David Boddy, *Secretary*

ACKNOWLEDGMENTS

The task committee wishes to thank two important groups for their assistance and contributions to this document. The corresponding members of the ASCE 74 committee provided substantial contributions based on their expertise in their respective fields. The corresponding members are

Kelly Bledsoe
Ahmed Hamada
Kathy Jones
Leon Kempner

The second group deserving much praise for their assistance and candid observations is the Peer Review Committee. It has been a pleasure to work with these individuals. Their contributions are greatly appreciated.

Ronald Randle, *Chair*
Anthony DiGioia
Eric Ho
Jon Kell
Jean-Pierre Marais
Robert Nickerson
Alain Peyrot
Tim Wachholz

CHAPTER 1
OVERVIEW OF TRANSMISSION LINE STRUCTURAL LOADING

1.0 INTRODUCTION

This Manual of Practice addresses structural loadings to be applied to transmission lines in the interest of reliable and cost-effective designs in compliance with regulations, standards, and prescribed design methods. The following key topics are addressed:

- Uniform procedures and definitions used in the industry for the calculation of loads. These are intended to facilitate consistency and communication in the transmission design industry.
- Design procedures that recommend a uniform level of reliability for transmission lines, as well as a means for increasing or decreasing this reliability when required. Depending on their importance, some transmission lines may justify the use of a greater level of reliability. These procedures may also be used to benchmark the reliability of existing lines.
- Procedures for calculating design loads and determining their corresponding load factors. Component and material strengths and strength factors must also be determined, although the scope of this manual is limited to general guidance. The designer is directed toward material-, component-, or product-specific references to obtain the values to be used with this methodology. Loading criteria should contain a comprehensive set of loads, as well as appropriate load factors associated with uncertainty. When properly coordinated with factored material strengths (which reflect the

variability of materials), the desired levels of reliability can be obtained. Reliability levels can then be adjusted to meet the performance needs and risk tolerance necessary for the facility.
- Failure containment philosophy for structures that are intended to reduce the probability of cascading failures.
- The most current techniques for quantifying weather-related loads, namely wind or ice and concurrent wind, that typically control the design of transmission line structures. High-intensity winds (HIWs) such as downbursts and tornadoes are also addressed.
- Explanations of wire systems, including how wire tensions and loads affect a transmission line system.
- Practical examples giving more detail on the application of load recommendations.
- Appendices containing background information and detailed discussion of several aspects of transmission line design. This additional information is intended to supplement the procedures presented in the main chapters of the manual.
- An appendix containing a draft Pre-Standard for design loads for electrical transmission line facilities.

Because this Manual of Practice is considered a guideline, it represents the most current and relevant loading concepts and applications specific to transmission line design.

1.1 PRINCIPAL SYSTEMS OF A TRANSMISSION LINE

A transmission line consists of two distinct structural systems: the structural support system and the wire system. The structural support and wire systems are often considered separately, although they are joined and respond to loading as a single system. Characteristics and roles of the structural support and wire systems are described in the following sections.

1.1.1 Wire System

The wire system consists of the conductors and overhead ground/shield wires and includes all components such as insulators and other hardware used to attach the wires to the support structures. The majority of the loading on transmission line structures is attributed to the wire system under gravity and environmental loads (i.e., wind and combined ice and wind). In addition, much of the unusual behavior and loading challenges faced by the line designer are generated by the wire system (i.e., unbalanced tensions, wire vibration, galloping, and broken wire impacts). It is therefore critical to have a thorough understanding of the wire system to understand the loading and response of the overall transmission line system. Loads applied to support structures are typically described in relation to the alignment of the transmission line. Longitudinal loads are applied to the structure in parallel to the transmission line and are caused by unequal wire tensions on adjacent spans, wire termination, or wire/tower failure. Transverse loads are applied normally to the transmission line and result from wind on the wire system (either bare or ice-covered wires) plus the wire tension resultant because of a line angle (if any). Vertical loads are due to the self-weight of the wires and attachments, the vertical component of wire tension, and any accumulated ice.

1.1.2 Structural Support System

The structural support system consists of the towers, poles, guys, and foundations; and supports the load from the wires, insulators, hardware, and wire attachments. It also resists wind and combined ice and wind loads on the wire system, as well as wind and ice loads on the structural components. The structural support system, or structure, is an essential element of a transmission line. Each structure is usually evaluated separately, although they are joined by the wires, which can transfer load and cause them to act as a single system. Consideration can be given to evaluating the structure as part of a system and considering the resisting effects of the wires that join them.

Structures are typically classified based on the function they perform in supporting the wire system. The following categories of structures are commonly referred to in transmission line design, and are referred to throughout this manual:

- Tangent structures: Structures that primarily resist the vertical weight of the wire (and accumulated ice) and transverse loads due to wind on the structure and wire system. Structures having small line angles (generally less than 2°) are commonly referred to as tangents. Although tangent structures can resist substantial transverse and vertical loads at relatively low cost, these structures may not provide adequate resistance in the case of a high longitudinal load. Some tangent structure configurations are less resistant to high longitudinal loads. For these structures an unanticipated imbalance in longitudinal loading, such as a broken wire, may initiate a cascade failure.
- Angle or Running Angle structures: Structures marking changes in line angle along the transmission line where the wires are held in suspension. Angle structures support the same vertical and transverse loads as tangent structures, as well as significant transverse loads resulting from wire tensions being applied at an angle.

- Dead-end structures: Structures designed as termination points for wires in one or more directions. Although the majority of these structures under everyday conditions will have wires spanning into both adjacent spans, they may at some point, usually during construction and possibly as a result of some type of failure event in extreme weather conditions, have full design wire tensions applied on one side (ahead or back span) only. As such, these structures are often designed to support the full design tension of all wires pulling longitudinally on one side (ahead or back span) or face of the structure.
- Strain structures: Structures including both angle strain structures and in-line strain structures. Structures similar to dead-end structures, where wire tensions are transferred directly to the structure but are capable of resisting only unbalanced/differential tensions. These structures may be used for a variety of reasons, including supporting tension during wire stringing, accommodation of clearance limitations, prevention of cascade failure, and resisting uplift conditions resulting from differences between adjacent spans (e.g., span length, slope, design tension).

1.2 UNIQUE ASPECTS OF TRANSMISSION LINE DESIGN

1.2.1 Tolerance of Failure

A unique aspect of structural design of electrical transmission line facilities is that failure at some level is acceptable. The acceptable level of risk of failure often depends on the importance of the transmission line considered. Transmission grids typically have some level of service redundancy, which can accommodate failure of a particular transmission line without any disruption of service. In some cases, a component of a transmission line can fail and may only damage a small portion of a line, which can be promptly repaired and service restored with minimal impact to the electrical grid. This is unique in comparison to other engineered structures (e.g., buildings, bridges, dams), where a failure could directly result in high probability of loss of life or substantial property damage. Engineering judgment should be used to balance reliability of design, minimize the probability and extent of failure, and provide economical design for the service life of the transmission line.

There are exceptions where failures resulting in a disruption of service are to be avoided. Some transmission lines serve critical facilities (e.g., hospitals, emergency services, power plants, cold-start facilities), which may be in congested, heavily populated areas, or may not have redundancy in the grid such as a critical radial feed line. It is recommended to design these critical transmission lines to a higher level of reliability to reduce the probability of failure. The designer is directed to Appendix L or ASCE 7-16 (2017), *Minimum Design Loads and Associated Criteria for Buildings and Other Structures*, for additional weather loading data for higher MRIs.

1.2.2 Designing to Contain Failure

Recognizing that failure of transmission line components may occur, it is important to consider failure mitigation or containment in the design of the line. This failure mitigation or containment is most often addressed by the use of structures designed for failure containment loads inserted at regular intervals along the line. This is covered in more detail throughout the document in the discussion of failure containment. Failure containment is important to consider in order to prevent the devastating effects of cascading failures, which could result in almost complete destruction of all, or major portions, of a transmission line.

1.2.3 Coordination of Strengths

The fundamental purpose of coordinating strengths of components or groupings of components is to limit and contain damage from unanticipated loadings in such a way that repairs needed to restore a damaged transmission line would be faster and more economical. As such, transmission line systems can be designed in such a way that failure will first occur in a component which is easy to replace, before other components or the entire structure, which are more difficult to replace, are damaged. For example, a tower arm or steel pole arm may be designed to fail at a load less than that which would cause the entire structure to collapse.

Designing for a sequence of failure through coordination of component strengths is often difficult as the actual performance may not meet expectations. This is often due to the large variability in the strength and strength distributions of the various components of a transmission line. The actual failure point of any individual component is difficult to predict with accuracy, especially when considering the different rates at which materials deform and deteriorate over time. Many components are also manufactured for nominal strength ratings (e.g., 15,000, 25,000, 36,000, 50,000 lb), which often results in some components having a greater capacity than would otherwise be required. Despite the difficulties in achieving coordination of strengths, it is good design philosophy for the designer to consider how and through what mechanisms their transmission line could potentially fail, and the consequences resulting from each failure.

1.2.4 Linear Exposure of Transmission Lines

The fact that transmission lines traverse long distances results in longer lines generally having a greater possibility of exposure to any low-probability events. This spatial characteristic of transmission lines results in longer lines having a lower inherent reliability and an increased probability of failure compared to shorter ones for the following reasons:

- Uncertainty resulting from exposure to differing terrain, land use, and natural and artificially manufactured crossings.
- Greater probability of exposure to a weather event striking or microclimatic condition occurring anywhere along the line. In addition, the loads produced by weather events are inherently spatially distributed and can act on transmission lines in singular or multiple locations with varying intensity over long distances (e.g., hurricanes and high-intensity winds).
- Greater probability that a weak component will experience an extreme loading event, possibly resulting in failure, because there are more components involved.

By recognizing this spatial characteristic of transmission lines, the designer must realize that the reliability of a long line is less than that of a shorter one, all other design parameters being equal.

Having noted the preceding points, it is also evident that it is difficult to select the appropriate load criteria based on the length of a given transmission line or line section. The result may be that structures and components suitable for a specified line length are not suitable for a line of a different length. One approach is to break long lines into loading zones, which allows the designer to maintain a consistent reliability level. Another approach is to establish standards that apply a greater reliability loading criteria to lines longer than a given length.

1.3 LOAD AND RESISTANCE FACTOR DESIGN (LRFD)

Although this Manual of Practice is a loading manual and not a design document, the load and resistance factor design (LRFD) concept is presented to provide context for the recommended loads. The relationship of load and load factors is presented in relation to various limit states that may need to be verified in the design or analysis of a transmission line.

1.3.1 Reliability-Based Design

Reliability-based design (RBD) procedures are used to set an acceptable level of probability of failure and design various components to satisfy this target reliability. In terms of statistics, this is achieved by limiting the overlap of the lower tail of the probability distribution of the component strength with the upper tail of the probability distribution of the load or load effect. However, to carry out a design using RBD procedures, a significant amount of material testing and weather data is required to accurately determine the probability distributions of the component strength and loading. This information is often not readily available. The increased level of input required, uncertainty in the transfer functions used to calculate loading from weather data, and the resulting loss of accuracy in calculating the probability of failure of a single component, let alone an entire system, seldom justify complex RBD procedures. Rather, an attempt is made to identify and articulate the most useful portions of RBD. The designer is directed to ASCE Manual of Practice No. 111, *Reliability-based Design of Utility Pole Structures* (2006) for additional discussion of RBD. The practical application of RBD continues to be focused on what can effectively be calculated with statistical weather and strength information in transmission line design work.

Two key elements of the RBD methodology are (1) methods for adjusting the relative reliability of a line design with respect to ice and/or wind loads, and (2) techniques for ensuring that structural components have appropriate strength levels relative to each other. These two elements are fundamental to maintaining control over the behavior of a line under extreme loading conditions.

1.3.2 Overview of LRFD

Load and resistance factor design methodology is a simplified approach to RBD which uses factors to account for uncertainty in the loading and strength (or other limiting condition). Through this method, a nominal strength and its corresponding strength reduction factor (ϕ, sometimes equal to 1.0) are used to estimate the statistical probability of failure and determine a nominal design capacity for the component. Conversely, loads and load effects are increased by load factors (γ) that account for the uncertainty and statistical variability of the load, and are used to establish design loads. Load factors are typically larger than 1, with occasional exceptions. Each component (i.e., dead load, wind load, construction load) of an applied load may have a different load factor representing the variability of that particular load. For example, the load factor for dead loads is typically much less than the load factor for wind loads due to the relative levels of uncertainty associated with each type of load. RBD methods are often used to calibrate appropriate strength reduction factor (ϕ) and load factors (γ).

Equation (1-1) shows the basic LRFD concept; in essence, strength must be greater than, or equal to, load. This relationship should be checked for all components of the system with their corresponding strength factors and limit states, and for all the various load cases and limit states.

$$\phi \cdot R_n \geq \sum \gamma_i \text{Load}_i \qquad (1\text{-}1)$$

where

ϕ = Strength reduction factor (a different strength factor may be used for each component and for each limit state);
R_n = Nominal strength of the component;
Load = The appropriate combination of dead loads, wire tensions, and weather-induced loads, construction and maintenance (C&M) loads, failure containment, and legislated loads; and
γ = Load factor, which is unique for each load.

1.3.3 Load Factors

Adjustment of the load factor can be applied to the following situations:

- To adjust the reliability with respect to a given weather-related load of a specific MRI,
- To account for uncertainty or unknowns in the load predictions, and
- To adjust for the consequences of failure such as worker safety.

For weather-related loads, load factor adjustments should not be combined with MRI adjustments without thorough evaluation. The use of load factors is appropriate for all non-weather-related loads.

1.3.4 Strength Factors

The purpose of the strength factor, ϕ in Equation (1-1), is to account for uncertainty in the coefficient of variation of the strength, COV_R, and the nonuniformity of exclusion limits that currently exist in published formulas for nominal strength, R_n. Different strength factors are also used for different limit states; for example, different factors would be used when checking the ultimate strength and the cracking limit of a component. The objective of the line designer is to apply appropriate strength factors to each component, or groups of components, to control the reliability (or probability of failure) of the transmission line.

The development or determination of strength factors for components is beyond the scope of this manual. Guidance on the determination and use of strength factors is available in IEC 60826, ASCE Manual of Practice No. 111 (2006), material design standards and guides, and manufacturer specifications.

1.3.5 Sources for Nominal Strengths

The following are a selection of standards and guidelines that may be used to develop nominal strengths (and strength factors) for components commonly found in transmission lines:

- ASCE 10, *Design of Latticed Steel Transmission Structures;*
- ASCE 48, *Design of Steel Transmission Pole Structures;*
- ASCE 104, *Recommended Practice for Fiber-Reinforced Polymer Products for Overhead Utility Line Structures;*
- ASCE Manual of Practice No. 123, *Prestressed Concrete Transmission Pole Structures;*
- ASCE Manual of Practice No. 141, *Wood Pole Structures for Electrical Transmission Lines;*
- ANSI Standards O5.1, O5.2, and O5.3 for wood poles, laminated timber, wood cross-arms, and braces;
- ANSI C29 Standards for both ceramic and nonceramic insulators;
- ANSI C119 Standards for conductor connectors including dead-end connections;
- IEEE specifications for various pole line hardware;
- ACI 318, *Building Code Requirements for Structural Concrete;* and
- AISC 360, *Specification for Structural Steel Buildings.*

1.3.6 Limit States

While ultimate strength design and other limiting factors are beyond the scope of this document, it is important to recognize that various limiting criteria, from deflection to ultimate strength, may control the structural design of a transmission line. It is therefore important to select the appropriate loads for the limit under consideration:

- *Failure limit* is the point at which a component of the line can no longer sustain or resist the imposed load. Exceeding this limit will likely result in the failure of some portion of the line.
- *Damage limit* is the point at which a component of the line will suffer permanent damage but may still function, possibly at a reduced level. Any permanent damage experienced (i.e., plastic deformation) may affect the future performance and serviceability of the line. It may also result in a reduced capacity to handle loads at the damage limit or failure limit levels. Examples of this include overstressed conductors that may need to be resagged, or hardware fittings that may be deformed to the point where maintenance is difficult or impossible. Insulators that have been loaded beyond their recognized safe working values, and guys which have been overstressed and require retensioning are also common examples.

Conventional practice is to ensure that under weather-related loading, and during construction and maintenance (C&M) operations, the uses of these components are kept below their damage limit. This is typically taken as a percentage of their rated (or nominal) strength. To do this effectively, it is important to understand how the rated or nominal strength is defined by the manufacturer.

- *Deflection or serviceability limits* are established to meet a variety of needs: pole top deflection may be limited for aesthetic purposes; wire blowout must be checked for various loading conditions to ensure proper clearances are maintained; changes in conductor sag and the possible impaired clearances resulting from structure deflections must be considered; and foundation deflections are often limited to ensure that excessive nonrecoverable deformations are not experienced. Due to the nonlinearity of material properties and the relationship between load and deflection, it is recommended to determine these deflection or serviceability limits using nonfactored service load cases having a high probability of occurrence during the expected life of a transmission line.

1.4 WEATHER-RELATED LOADS

Weather-related loads on transmission line structures and wires are associated with wind, ice, and temperature or a combination of those loads. Atmospheric pressure and local topography can influence the characteristics of weather-related loads; these influences should be considered where appropriate, based on engineering judgment, expert opinion, and past practice experience.

1.4.1 Extreme Winds

The methodology for the calculation of wind loads on transmission structures, components, and wires is presented in Chapter 2. The extreme wind speeds recommended for design are primarily based on the provisions of ASCE 7-16. The general extreme wind loading methodology is associated with synoptic wind events, including hurricanes.

1.4.2 High-Intensity Winds

Wind loads resulting from convective events, such as thunderstorms, downbursts, and tornadoes, differ in magnitude, load distribution, and duration from the extreme winds that transmission structures are normally designed to resist. Damage and failure resulting from HIWs have been frequently described for many locations around the world. Some locations experience more of these events than others, and engineering judgment and past experience with the area should be considered when applying HIW methods. A general description of HIW is given in Chapter 2, while approaches to design for HIW (including downbursts and tornadoes) are given in Appendix K.

1.4.3 Extreme Ice with Concurrent Wind

Extreme ice with concurrent wind speeds are presented in Chapter 2. The formation of ice on transmission structures is generally referred to as ice accretion and is often more important for the wires than for the structures. Ice accretion on a transmission line often imposes substantial vertical loads on the structural system. The resultant load on the wires causes significantly greater wire tensions compared to bare wire conditions. Chapter 2 provides a basis for estimating the thickness of ice on conductors and shield wires, as well as the concurrent wind speeds and air temperatures to be considered with ice accretion.

1.5 RELIABILITY CONCEPTS FOR WEATHER-RELATED LOADS

1.5.1 Mean Recurrence Intervals for Weather-Related Loads

It is customary to associate extreme values of weather events with some MRI, which is commonly referred to as a return period. For example, a MRI of 100 years is associated with an event which, on average, has a probability of 1/100 (or 1%) of being met or exceeded in any given year or a 39% chance of being met or exceeded at least once during a 50-year period. However, because extreme weather events are not necessarily evenly spaced over time, some 50-year periods may pass with no weather equal to or exceeding the level associated with the 100-year MRI. Conversely, some 50-year periods may experience multiple events equal to or exceeding the level associated with the 100-year MRI. The MRI analysis assumes that the probability of the event occurring does not vary over time and is independent of past events. The following formula can be used to calculate the probability of exceeding the level corresponding to a given MRI over an N-year period.

$$\text{Probability of Exceedance in } N \text{ years} = \left[1 - \left(1 - \frac{1}{MRI}\right)^{N}\right] \qquad (1\text{-}2)$$

Table 1-1 includes a column showing the probability of exceedance of various MRIs one or more times during a span of 50 years. For example, the probability of exceeding the wind speed associated with a 50-year MRI

Table 1-1. Exceedance Probability for Various MRIs.

Typical conditions	MRI (years)	Probability the MRI load is exceeded in any one year (%)	Probability the MRI load is exceeded at least once in 50 years (%)	Probability the MRI load is exceeded at least once in 100 years (%)	Location of wind and ice maps shown in ASCE MOP 74	
					Wind	Ice
Temporary or emergency restoration, service checks[a]	10	10	99	99+	[e]	—
Temporary or emergency restoration, service checks[a]	25	4	87	98	[e]	—
Historically used MRI[c]	50	2	64	87	Appendix L	Appendix L
Recommended MRI	100	1	39	63	Chapter 2	Chapter 2
Enhanced reliability	200	0.50	22	39	—	—
Enhanced reliability	300	0.30	15	28	Appendix L	Appendix L
Enhanced reliability	400	0.25	12	22	—	—
Enhanced reliability	500	0.20	10	18	—	[d]
Enhanced reliability	700	0.14	7	13	[b]	—

[a]Service checks include clearance checks, blowout checks, insulator uplift, and deflections.

[b]ASCE 7-16, Chapter 26.

[c]Previous editions of this manual provided wind speeds associated with a 50-year MRI for a majority of the continental United States, with MRIs of the hurricane-prone regions in the range of 50 to 90 years. See Section 1.5.1 and Chapter 2 for further explanation.

[d]ASCE 7-16, Chapter 10.

[e]ASCE 7-16, Appendix C Commentary.

at least once in a 50-year period is $[1-(1-(1/50))^{50}] = 0.64$, or 64%. It should also be noted that there is an almost 1 in 4 probability of a 200-year MRI event occurring at least once in 50 years, $[1-(1-(1/200))^{50}] = 0.22$, or 22%.

Maps of extreme wind and extreme ice with concurrent wind are presented in Chapter 2. The extreme wind and ice loads are provided for specific MRIs. Applying loads for a different MRI can be used to vary the target reliability of the design. For weather-related loads, the recommended method to adjust the relative reliability level is to use a load corresponding to the MRI of interest. This is most accurately done by using the extreme wind and extreme ice with concurrent wind maps provided, as the relationship between extreme weather-related loads and MRI varies spatially throughout the United States. The use of relative reliability load factors to transform basic wind speeds and ice thicknesses to different MRIs is not promoted in this edition. Rather, maps relating to specific MRIs have been proposed to improve consistency of risk, as detailed in Chapter 2.

This edition of the manual recommends using 100-year MRI extreme wind and ice loads as the basis for design. Extreme wind and ice maps for additional MRIs are provided in Appendix L and ASCE 7-16.

The previous edition of this manual presented a basic wind speed map, taken from ASCE 7-05 (2005), which was commonly interpreted as a 50-year return period wind speed map. However, the return period varied in hurricane-prone regions from 50 to 90 years. Further information is provided in the commentaries to the wind load chapters in ASCE 7-05 and ASCE 7-10 (2010).

The wind speed maps in this edition of this manual are adopted from ASCE 7-16. The maps in ASCE 7-16 address the issue of risk consistency between nonhurricane and hurricane regions and incorporate additional wind data obtained since the development of the previous map. The 100-year MRI wind speed map in Chapter 2 provides the same risk consistency and reliability across the United States.

1.5.2 Relative Reliability and Weather Event MRIs

The reliability of a transmission line, a single structure, or a single component can be adjusted up or down by choosing higher or lower MRIs, respectively. Several factors should be considered prior to adjusting the design MRI for a transmission line or one of its components. For example, considering the critical importance of the facilities served, or the generation source, the line owner may desire an increased level of reliability and therefore select design loads associated with a higher MRI. Unique structures that may be difficult to replace, structures at major road or river crossings, or structures in a high-density urban area may also justify an increased level of reliability.

Conversely, for temporary lines, emergency lines, existing transmission lines with relatively short life spans, or where the consequences of failure are deemed relatively inconsequential, a decreased level of reliability may be deemed appropriate. The applicability and requirements of legislated loads must be considered when reducing the MRI as to not violate any minimum requirements. Sound engineering judgment should always be used when deviating from the 100-year MRI recommended in this Manual of Practice.

The relative probability of exceedance of a load is inversely proportional to the design load MRI (Peyrot and Dagher 1984). Hence, doubling the MRI of the design load reduces the probability of exceedance in any one year by a factor of 2, as shown in Table 1-1.

The concept of relative reliability can be used as a tool to approximately adjust design reliability as it is currently very difficult to accurately calculate the probability of failure of a line. Powerful mathematical tools which can predict transmission line reliability are available, but such detail is often not justified considering the linear extent of a line, the redundancy normally included in transmission systems, and the availability of the data required to perform such an analysis. Specifically, predicted probabilities of failure are in error if the uncertainties in the probability distribution of climatic events, structure strengths, component strength, and transfer functions converting climatic events into loads are not adequately considered. Until more information becomes available to resolve the load and strength data issues, it is recommended that the concept of relative reliability be used, as an admittedly approximate tool, to adjust structural reliability.

1.5.3 Service Reliability versus Structural Reliability

There are many events that affect the service reliability of a transmission line. Some are related to electrical events and may be controlled by station equipment rather than transmission line design. Others may be storm-related, such as electrical flashovers experienced during lightning events. These aspects are not included in the scope of this manual but contribute to the overall electric service reliability of the transmission line. When determining the desired service reliability, one must separate the electrical service reliability from the structural reliability.

1.6 ADDITIONAL LOAD CONSIDERATIONS

There are several other types of loads, loading conditions, and information on the time signature of loading (see Section 1.6.5) that may need to be considered in the design or analysis of a transmission line and its components. Chapter 3 includes a detailed discussion of some of these other types of loading, including construction and maintenance loads, and longitudinal and failure containment loads. Chapter 3 also includes discussions regarding load effects due to galloping wires, as well as potential problems due to the effects of vibration of structural members and seismic events.

1.6.1 Construction and Maintenance

Construction loads are loads the structure must resist during assembly, erection, and throughout the installation of shield wires, insulators, conductors, and line hardware.

Maintenance loads are experienced by the structure as a result of inspection, replacement of structure components, or alteration of the supported facilities. Structure maintenance loads consider the effects of workers on the structure, as well as load effects on adjacent structures due to temporary modifications such as guying to repair or replace a structure. Personnel safety should be the paramount factor when establishing all loads, especially C&M loads. Construction and maintenance loads and conditions are discussed in more detail in Chapter 3.

1.6.2 Longitudinal and Failure Containment Loads

Structures may experience longitudinal loads (as a result of tension differences in adjacent spans) in a variety of scenarios. These loads should be included in the design of structures when warranted. These scenarios include, but are not limited to

- Inequalities of wind and/or ice on adjacent spans (e.g., ice on the back span with no ice on the ahead span),
- Ice- and/or wind-load-induced longitudinal imbalance on consecutive spans with significant difference in length (e.g., tangent span over a valley),
- Wire breakage,
- Construction or maintenance loading scenarios (e.g., wire caught in the stringing block),
- Hardware swing constraints that restrict balancing of tensions,
- Insulator failure, and
- Structural and hardware component failure.

Longitudinal loading events can create severe load imbalances in the wire system capable of causing the partial or complete failure of the adjacent supports. Propagation of the failure to a multitude of support structures along a transmission line as a result of longitudinal load imbalances is generally referred to as a cascade failure. These cascade failures cause significant damage and high economic losses as complete sections of a transmission line may be destroyed, resulting in weeks or months of repair (Ostendorp 1997). Cascade-type failure events can be prevented or minimized by the inclusion of certain longitudinal load cases and scenarios. Longitudinal loads and approaches to failure containment are discussed in further detail in Chapter 3.

1.6.3 Earthquake Loads

Historically, transmission line structures have not failed due to inertial loads caused by earthquakes. Decades of industry experience and records support the observation that little, if any, damage is observed on transmission structures after an earthquake event. Therefore, the structural design of transmission line structures need not consider inertial loads associated with earthquakes.

While the design of transmission line structures will not be controlled by earthquake-induced inertial loads, structural damage could occur as a result of secondary effects, such as damaged foundations or tiebacks, use of rigid post-type insulators, minimal wire slack between structures, soil liquefaction and landslides. When locating or designing a new transmission line route, recognition of various seismic hazards should be considered to minimize potential damage. Additional descriptions of hazards and potential secondary effects due to ground motion are provided in Chapter 3.

1.6.4 Legislated Loads

Legislated loads refer to minimum loads specified by federal, state, or municipal codes and legislative or administrative acts. In the United States, the most common source of legislated loads is the National Electrical Safety Code (NESC) (IEEE 2017), which specifies minimum loading requirements for safety as specified in Rule 010.

Legislated codes may require load cases in addition to those imposed by atmospheric conditions, such as those caused by anticipated construction and maintenance activity. In addition to loads, allowable material strengths, load factors, and calculation methodologies may also be specified. Legislated loads are not included in the scope of this manual.

1.6.5 Load Time Signature

When prescribing a load for a given situation, information in addition to the magnitude is required to achieve the desired transmission line performance. The designer may also need to consider the time history of the load, or load signature, as the resistance of materials can vary with load duration, dynamic load characteristics, and load intervals (impact or cyclical).

The following examples illustrate the response of different materials to various loading characteristics:

- Wood, nonceramic insulator, and fiber-reinforced composite member strengths must be reduced when exposed to long-duration loads, such as those that might occur during an icing event or recurring cumulative load events;
- Metallic materials exhibit fatigue when exposed to cyclical loading events such as vibration;
- Impact loads, such as those which may occur during a conductor galloping event, can be detrimental to ceramic materials; and
- Hardware and connections can be loosened by vibration.

As can be seen from this list, additional checks may be required depending on the load signature and affected component materials.

1.7 WIRE SYSTEM

To determine the loads on transmission structures accurately, it is necessary to understand how tensions are generated in the wire systems, and how the resulting loads are imposed on the support systems. These aspects are described and discussed in Chapter 4. Wire tension changes with temperature, time (creep), ice, wind, the flexibility of the support, and C&M operations. Chapter 4 discusses the manner in which the wire system responds to these unit wire loads and some of the assumptions that may be used for determining the loads at the structure attachment points. Chapter 4 also provides information regarding

- Effects of creep and heavy loads on wire tensions,
- The need to keep wire tensions within certain limits, and
- Simplifications and assumptions that may be used for determining wire tensions and resulting loads at structure attachments.

1.8 EXAMPLES

The application of the recommended loadings in Chapters 2 and 3 is illustrated for a typical suspension tower in Chapter 5. Appendix F contains detailed example calculations for the gust response factors for the same tower.

1.9 APPENDIXES

A number of appendixes are included, which provide additional information to supplement the various loading topics covered in the main body of the manual. Where applicable, the relevant appendix is referenced in the corresponding chapter of the manual. Additional information, background and history, and derivations are available for the majority of the recommended loading provisions.

1.10 DRAFT PRESTANDARD

One of the goals of this edition of the manual was to create a draft presentation for a standard for Transmission Line Loading. The ultimate goal is to develop a standard for Design Loads for Electrical Transmission Line Facilities. The initial draft of this prestandard document is included in Appendix M in order to disseminate it for public review and comment, and to gather and consider industry consensus for future revision and publication.

1.11 INCORPORATION OF CHANGING DATA

When applying this Manual of Practice to design, the line designer must recognize that the contents of this manual were developed based on historical performance of both environmental loading as derived from recorded meteorological data and of the performance of transmission line structures in response to this loading. In other words, the knowledge accumulated and implemented in this and other manuals and guidelines is based on data and performance of the past. The database of weather records is continuously expanding with each passing year and as more weather stations are installed. The importance of obtaining new meteorological and performance data is more important with global climate change, as the effects of these changes on the electrical system are largely unknown. Proactive efforts to gather and apply additional data should be engaged where possible and practical to ensure electrical service continuity relative to structural loading.

CHAPTER 2
WEATHER-RELATED LOADS

2.0 INTRODUCTION

This chapter discusses weather-related loads on transmission line structures and wires. These loads are associated with wind or a combination of ice and wind, referred to as ice with concurrent wind. Temperature, atmospheric pressure, and local topography influence the magnitude of weather-related loads. These influences should be considered where appropriate.

A standard wind pressure formula applicable to transmission lines is presented. The formulation of design wind pressures recommended in this chapter is primarily based on the provisions of ASCE 7-16 (2017). The wind equations presented in this Manual of Practice are developed from information currently available in the engineering community and provide practical methods for design resulting in adequate levels of performance. However, due to several challenges, as well as the spatial extent of transmission lines, precise wind load prediction is difficult. Some of these challenges include the uncertainty in quality and quantity of wind data, the transfer functions to convert wind velocity to wind pressure, and the various meteorological mechanisms generating extreme wind (e.g., large-scale pressure systems, convective storms). It has been well documented that the load effects resulting from convective events, such as thunderstorms, downbursts, and tornadoes, differ from those for which transmission structures are normally designed. These types of winds are often referred to as high intensity winds (HIWs).

Ice with concurrent wind loads are also described. The formation of ice on transmission line components is generally referred to as ice accretion. This manual provides regional values of ice thickness due to freezing rain and the wind speeds and air temperatures to be considered in combination with ice accretion.

Supplemental information on converting wind speed to pressure, wind speed average times, gust response factors, force coefficients, and ice accretion is given in Appendixes C, D, F, G, and H, respectively. Appendix K provides advice and methods regarding the consideration of HIWs in the design of transmission structures. Appendix L provides weather-related loads for additional MRIs which may be of interest in design.

2.1 WIND LOADING

2.1.1 Wind Force

The wind force acting on the surface of transmission line components can be determined by using the wind force expression shown in Equation (2-1)

$$F = QK_zK_{zt}(V_{100})^2GC_fA \tag{2-1a}$$

or

$$F = QK_zK_{zt}(V_{\text{MRI}})^2GC_fA \tag{2-1b}$$

where

F = Wind force in the direction of wind unless otherwise specified [lb (N)];
G = Gust response factor for conductors, ground wires, and structures as specified in Section 2.1.5;
C_f = Force coefficient values as recommended in Section 2.1.6;
A = Area projected on the plane normal to the wind direction [ft^2 (m^2)];
Q = Air density coefficient defined in Section 2.1.2;
K_z = Wind pressure exposure coefficient, which modifies the reference wind pressure for various heights above ground based on different exposure categories defined in Section 2.1.4. The values are obtained from Equation (2-3) or Table 2-2;
K_{zt} = Topographic factor obtained from Equation (2-16) in Section 2.1.7;
V_{100} = Reference 3-second gust wind speed for 100-year MRI [mph (m/s)], obtained from Figure 2-1 in Section 2.1.3; and
V_{MRI} = Reference 3-second gust wind speed for selected MRIs [mph (m/s)], obtained from ASCE 7–16 or Appendix L of this manual.

The wind force calculated from Equation (2-1) is based on the selection of appropriate values of wind speed, wind pressure exposure coefficient, gust response factor, and force coefficient. These parameters are discussed in subsequent sections. Wire tension corresponding to the wind loading should be calculated using the wire temperature that is most likely to occur at the time of the extreme wind loading event.

2.1.2 Air Density Coefficient, Q

The air density coefficient, Q, converts the kinetic energy of moving air into potential energy of pressure. For wind speed [mph (m/s)] and pressure [psf (Pa)], the recommended value is

$$Q = 0.00256 \text{ mph to psf } (0.613 \text{ m/s to Pa}) \qquad (2\text{-}2)$$

The nominal value of Q in Equation (2-2) reflects the specific weight of air for a standard atmosphere [i.e., temperature of 59 °F (15 °C) and sea level pressure of 14.7 psi (101.325 kPa)]. For some cases, such as line upgrading/re-rating in high elevations, the effects of temperature and elevation (atmospheric pressure) on the value of Q may be considered. Sufficient meteorological data should be available to justify a different value of the air density coefficient for a specific design application. Variation of Q with temperature and elevation is discussed in Appendix C.

2.1.3 Basic Wind Speed

In the United States, the basic wind speed is the 3-second gust wind speed at 33 ft (10 m) above ground in open country terrain (Exposure Category C, as defined in Section 2.1.4.1). ASCE 7-16 provides basic wind speeds associated with various MRI in the form of contour maps. Wind speed maps for MRI of 10, 25, 50, 100, 300, 700, 1,700, and 3,000 years are provided in ASCE 7-16.

In the previous versions of this manual (ASCE 1991, 2010a), a single map of basic wind speeds was provided for the United States. Relative reliability factors were then provided to transform the basic wind speed, which was associated with a nominal 50-year MRI, to additional MRIs of interest (i.e., 25 years, 200 years). This process assumed that the relationship of wind speed and MRI is consistent among all regions. The study of wind climatology through longer historical records, as well as improvements in data collection and processing, has led to a different approach in the wind engineering community.

Based on an updated analysis of historical wind data in the United States, wind speed maps corresponding to several MRIs have been developed and are presented in ASCE 7-16 (e.g., Pintar et al. 2015). The wind speed maps included in the main body of ASCE 7-16 represent MRI associated with factored strength design (i.e., 300, 700, 1,700, 3,000 years) based on target reliability levels for building occupancy and function. Wind speed maps not associated with ASCE 7 strength factors for building occupancy and function (i.e., 50, 100 years) are available in Appendix C of ASCE 7-16.

The wind speed map recommended by this manual corresponds to a MRI of 100 years, consistent with the 100-year MRI wind speed map in ASCE 7-16, and is shown in Figure 2-1. This map provides wind speeds that are consistent with the intent of the basic wind speed map in the previous editions of this manual with respect to reliability. Values are nominal design 3-second gust wind speeds in miles per hour (m/s) at 33 ft (10 m) above ground in Exposure Category C terrain. Linear interpolation between contours is permitted. It is acceptable to use the last wind speed contour of the coastal area for islands and coastal areas beyond the last contour. Further examination is recommended for unusual wind conditions in areas identified as Special Wind Regions. Additional wind speed maps for MRIs of 50 years and 300 years are provided in Appendix L. Selection of a MRI of a value other than 100 years may be desirable for certain applications.

The entire state of Hawaii is defined as a Special Wind Region on the current wind speed maps. This is due to the extreme topographic conditions found throughout the Hawaiian Islands. Significant research on wind speedup due to topographic features has been carried out for the Hawaiian Islands (e.g., Chock and Cochran 2005), resulting in maps of the topographic factor K_{zt}. These detailed maps are publicly available through the Department of Accounting and General Services for the state of Hawaii. Following a review of the K_{zt} maps for Hawaii, the following wind speeds are recommended for a MRI of 100 years: (1) A wind speed of 105 mph (47 m/s) for regions indicated as $K_{zt} \le 1.5$, and (2), a wind speed of $\left(86\,\text{mph} \times \sqrt{K_{zt}}\right)$ or $\left(38\,\text{m/s} \times \sqrt{K_{zt}}\right)$ for regions indicated as $K_{zt} > 1.5$.

In certain regions of the country, such as mountainous terrain, topographical characteristics may cause significant variations in wind speed over short distances. See Section 2.1.7 for further discussions on topographic effects. These variations in wind speeds are dependent upon local effects and cannot be effectively shown on the maps. In addition, Special Wind Regions indicate that wind speeds in these regions may vary significantly from those values indicated on the map. In these cases, the designer should consult a meteorologist or wind engineer to establish a design wind speed.

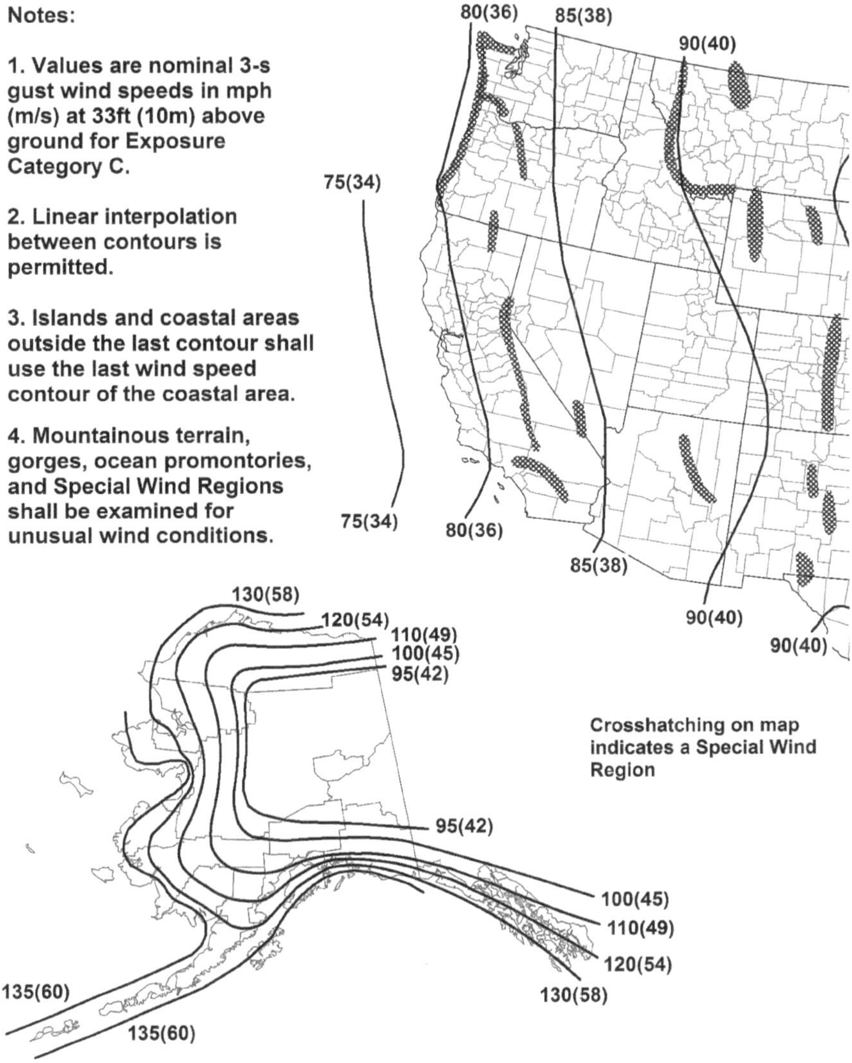

Figure 2-1. 100-year MRI 3-second gust wind speed map [mph (m/s)] at 33 ft (10 m) aboveground in Exposure Category C (Continued)
Source: ASCE (2017).

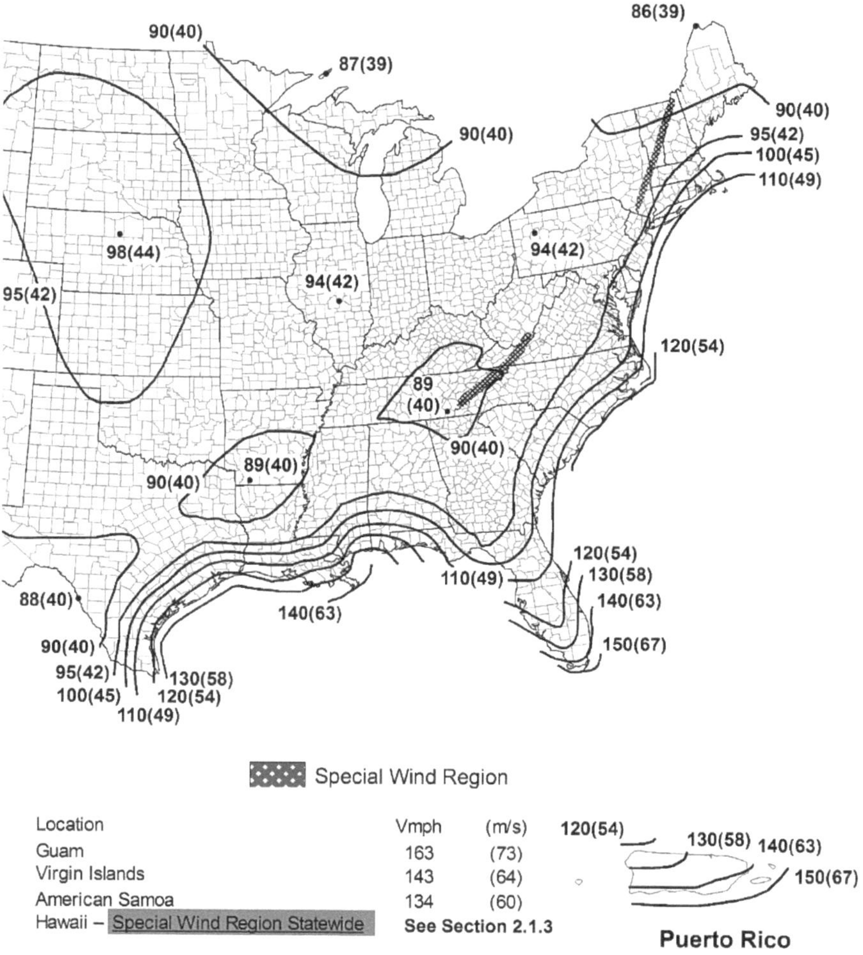

Figure 2-1. (Continued) 100-year MRI 3-second gust wind speed map [mph (m/s)] at 33 ft (10 m) aboveground in Exposure Category C
Source: ASCE (2017).

2.1.3.1 Use of Local Wind Data It is possible to determine the basic wind speed using regional wind data for a specific location. ASCE 7-16 provides criteria for the use of regional meteorological data.

2.1.3.2 Wind Speed Conversion A conversion procedure to obtain wind speeds of different averaging times is described in Appendix D.

2.1.4 Wind Pressure Exposure Coefficient

It is recognized that wind speed varies with height due to interaction (friction) with the surface of the earth. This is referred to as the atmospheric boundary layer and is dependent on ground surface roughness as characterized by the various exposure categories described in the next section. The wind pressure exposure coefficient, K_z, used in Equation (2-1) and defined in Equation (2-3) (Section 2.1.4.2), modifies the basic wind pressure, taking into account the effect of height and terrain.

2.1.4.1 Exposure Categories The following terrain roughness or exposure categories are recommended for use with this manual and are specified in ASCE 7-16. The recommended exposure category for transmission lines is Exposure C, unless the criteria of Exposures B or D can be met and there is a reasonable expectation that the exposure category will not change over the life of the transmission line. The entire state of Hawaii has detailed Exposure Category maps to correspond to the topographic maps described in Section 2.1. Use of these maps is recommended where applicable.

Exposure B This exposure is classified as urban and suburban terrain, densely wooded areas, or terrain with numerous, closely spaced obstructions having the size of single-family dwellings or larger. A typical view of terrain representative of Exposure B is shown in Figure 2-2. Use of Exposure B shall be limited to wind directions for which representative terrain

Figure 2-2. Typical terrain representative of Exposure B.

extends either 2,600 ft (792 m) or 20 times the height of the transmission structure, whichever is greater.

In the use of Exposure B, the question arises as to what is the longest distance of flat, unobstructed terrain located in the middle of a suburban area permitted before Exposure C must be used. A guideline is 600 ft (180 m) or 10 times the height of the transmission structure, whichever is less, as the length of intermediate, flat, open country allowed for continued use of the Exposure Category B. The surface conditions required for the use of the Exposure Category B are illustrated in Figure 2-3.

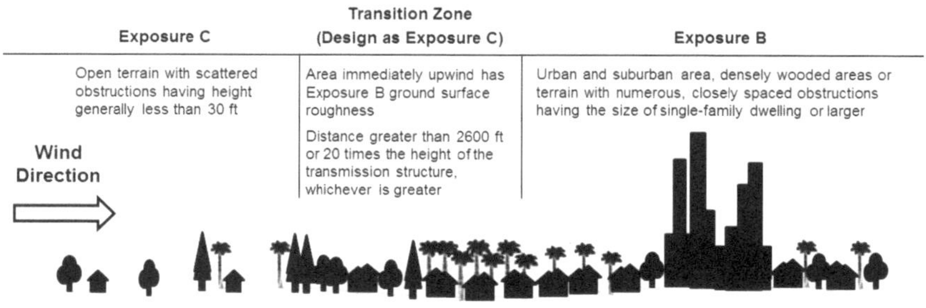

Figure 2-3. Surface conditions required for the use of Exposure Category B.

Exposure C This exposure is defined as open terrain with scattered obstructions having heights less than 30 ft (9.1 m). This category includes flat, open country, farms, and grasslands. A typical view of terrain representative of Exposure C is shown in Figure 2-4. This exposure category should be used whenever terrain does not fit the descriptions of the other exposure categories. It should also be noted that this exposure is representative of airport terrain, where most wind speed measurements are recorded.

Exposure D This exposure is described as flat, unobstructed areas directly exposed to wind flowing over open water for a distance of at least 5000 ft (1524 m) or 20 times the height of the transmission structure, whichever is greater. Note that the ASCE 7-16 wind speed map requires the use of Exposure D along the hurricane coastline (Vickery et al. 2010). A typical view of terrain representative of Exposure D is shown in Figure 2-5. Shorelines for which Exposure D also applies include inland waterways, the Great Lakes, and coastal areas of California, Oregon, Washington, and Alaska. The Exposure Category D applies only to structures directly exposed to bodies of water and coastal beaches. The surface conditions required for the use of Exposure Category D are illustrated in Figure 2-6.

Figure 2-4. Typical terrain representative of Exposure C.

Figure 2-5. Typical terrain representative of Exposure D.

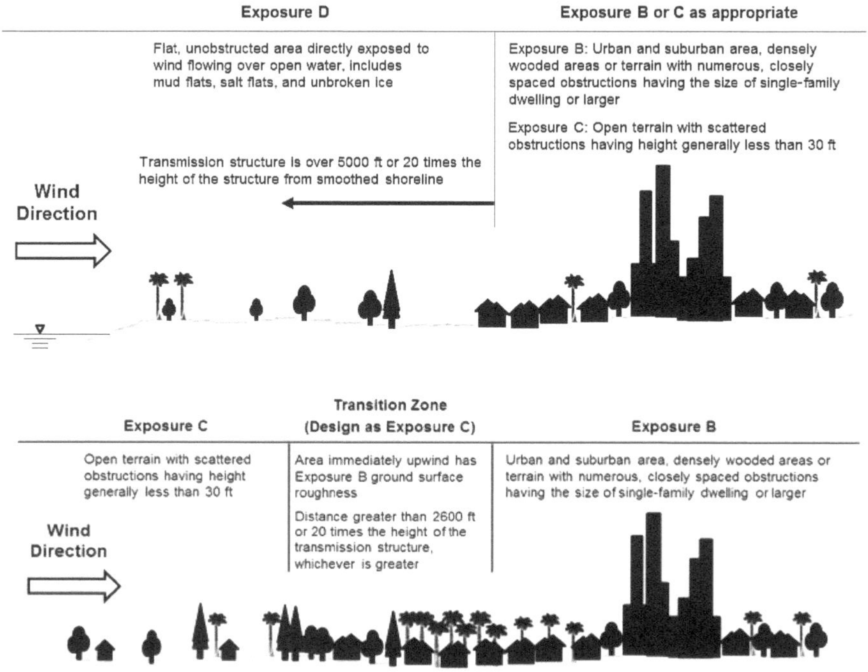

Figure 2-6. Surface conditions required for the use of Exposure Category D.

2.1.4.2 Exposure Coefficient Equations Values of the wind pressure exposure coefficient, Kz, are calculated using Equation (2-3).

$$K_z = 2.01\left(\frac{z_h}{z_g}\right)^{\frac{2}{\alpha}} \text{ for } 33 \text{ ft} \le z_h \le z_g \tag{2-3}$$

where

α = Power law coefficient for gust wind (see Table 2-1);
z_h = Effective height; and
z_g = Gradient height (see Table 2-1).

The effective height, z_h, is discussed in Section 2.1.4.3. The gradient height, z_g, defines the thickness of the atmospheric boundary layer. Above this elevation, the wind speed is assumed to be constant. The power law exponent, α, accounts for the shape of the wind speed profile with respect to height. Values for the power law exponent and corresponding gradient heights are listed in Table 2-1 for each exposure category. Table 2-2 can be used to determine K_z for heights up to 200 ft (60 m) above ground.

Table 2-1. Power Law Exponent for Gust Wind Speed and Corresponding Gradient Height

Exposure Category	α	z_g (ft)
B	7.0	1200
C	9.5	900
D	11.5	700

Table 2-2. Wind Pressure Exposure Coefficient, K_z.

Effective Height, z_h (ft)	Wind Pressure Exposure Coefficient, K_z		
	Exposure B	Exposure C	Exposure D
0–33	0.72	1.00	1.18
40	0.76	1.04	1.22
50	0.81	1.09	1.27
60	0.85	1.14	1.31
70	0.89	1.17	1.35
80	0.93	1.21	1.38
90	0.96	1.24	1.41
100	0.99	1.27	1.43
120	1.04	1.32	1.48
140	1.09	1.36	1.52
160	1.13	1.40	1.55
180	1.17	1.43	1.59
200	1.20	1.46	1.62

The effects of the wind pressure exposure coefficient on wind force for the different exposure categories are significant. It is essential that the appropriate exposure category be selected after careful review of the surrounding terrain. It is recommended that Exposure C be used unless the designer has absolutely determined that Exposure B or Exposure D is more appropriate. The transfer of the basic wind speed between exposure categories should only be used with sound engineering judgment.

2.1.4.3 Effective Height The effective height, z_h, is theoretically the height from ground level to the center of pressure of the wind load. The effective height is used for selection of a wind pressure exposure coefficient, K_z [based on Equation (2-3) or Table 2-2], and calculation of the gust response factors, G_w or G_t [Equations (2-4) and (2-5) in Section 2.1.5].

The effective height of a conductor and ground wire subjected to wind and ice with concurrent wind, is influenced by the blow-out swing of the wires and insulators. However, for structural design purposes, the effective heights of all the wires can be approximated as the average height above ground of all the wire attachment points to the structure.

The wind pressure exposure coefficient varies over the height of the structure. Transmission structures may be divided into sections, where the effective height, z_h, is the height to the center of each section. For some structures, a second or simpler alternative for structure heights 200 ft (60 m) or less is to assume one section and use two-thirds of the total structure height as the effective height. This alternative will apply a uniform wind pressure over the height of the structure. Sound engineering judgment should be used in the application of this approach.

2.1.5 Gust Response Factor

The gust response factor accounts for the dynamic effects and the correlation of gusts on the wind response of transmission line components. It is recognized that gusts generally do not envelop the entire span of wires between transmission structures, and that some reduction reflecting the spatial extent of gusts should be included in the calculation of the wind load. Both the dynamic effects and correlation have been incorporated in the gust response equations developed by Davenport (1979).

It should be noted that the gust response factor is different from the gust factor, which is used by some electric utilities in their wind loading criteria. The gust factor is the ratio of the gust wind speed at a specified short duration (e.g., 3 seconds) to a mean wind speed measured over a specified averaging time (e.g., hourly mean or 10- minute mean). The gust response factor is the ratio of the peak load effect on the structure or wires to the mean load effect corresponding to the design wind speed. Therefore, the gust factor is a multiplier of the mean wind speed to obtain the gust wind speed, whereas the gust response factor is a multiplier of the design wind load to obtain the peak load effect. The gust response factors described here replace the use of traditional gust factors.

The equations for gust response factors (Davenport 1979), described in Appendix F, were originally developed based on an hourly mean wind speed, as discussed by Behncke and Ho (2009). In the previous edition of this manual, the original Davenport gust response factors were modified to be consistent with the definition of the basic wind speed in ASCE 7, which is the 3-second gust at 33 ft in open country terrain (Exposure Category C). However, the equations retained parameters whose definitions have been improved upon since their original development. Thus, the equations for the gust response factors have been modified for this edition of the manual. Most notable are the removal of the parameters κ (surface drag coefficient) and E (exposure factor); these have been replaced with modern parameters used in the description of atmospheric boundary layer wind. The modified approach involves the calculation of the turbulence intensity of the wind at the effective height of the structure or wires, as well as the use of separate peak factors for the background and resonant components of the dynamic response. The revisions to the methodology reflect the state of the art in the calculation of wind loads on structures and are consistent with the methodology for atmospheric boundary layer winds applied in ASCE 7. The calculated wind pressure for Exposure Category C based on the updated GRF results in only a 3% difference (reduction) for a typical 100 ft structure compared to the previous method. The gust response factor equations are discussed further in this section, and an example of their application can be found in Appendix F. Further discussion on the updated gust response factors is provided by Mara (2015).

2.1.5.1 Equations and Nomenclature The gust response factor equations presented in this section have been simplified from the complete gust response factor equations described in Appendix F. That is, only the background component of the dynamic response is considered, which is likely sufficient for typical structures and span lengths. For tall or unique structures, or for longer span lengths, designers are encouraged to consider the complete gust response factors (including the resonance component) provided in Appendix F.

The simplified structure and wire gust response factors, G_t and G_w, respectively, are determined from the following equations:

$$G_t = \left(\frac{1 + 4.6 I_z B_t}{1 + 6.1 I_z} \right) \tag{2-4}$$

$$G_w = \left(\frac{1+4.6I_z B_w}{1+6.1I_z}\right) \tag{2-5}$$

in which

$$I_z = c_{exp}\left(\frac{33}{z_h}\right)^{\frac{1}{6}} \tag{2-6}$$

$$B_t = \sqrt{\frac{1}{1+\frac{0.56z_h}{L_s}}} \tag{2-7}$$

$$B_w = \sqrt{\frac{1}{1+\frac{0.8S}{L_s}}} \tag{2-8}$$

where

B_t = Dimensionless response term corresponding to the quasi-static background wind loading on the structure,
B_w = Dimensionless response term corresponding to the quasi-static background wind loading on the wires,
c_{exp} = Turbulence intensity constant, based on exposure (Table 2-3),
I_z = Turbulence intensity at effective height of the tower/structure or wire [Equation (2-6)],
L_s = Integral length scale of turbulence (ft) (Table 2-3),
S = Design wind span (ft) of the wires (conductors and ground wires), and
z_h = Two-thirds of the total height of the structure (ft) for the calculation of G_t [Equation (2-4)], or effective height of the wire (ft) for the calculation of G_w [Equation (2-5)].

Table 2-3. Turbulence Parameters for Calculation of Gust Response Factor by Exposure

Exposure	c_{exp}	L_s (ft)
B	0.3	170
C	0.2	220
D	0.15	250

2.1.5.2 Gust Response Factor for Structure

The structure gust response factor, G_t, is used in Equation (2-1) for computing the wind loads acting on transmission structures, structural components, and on the insulator and hardware assemblies attached to the structures. G_t, given by Equation (2-4), is a function of the exposure category (defined in Section 2.1.4.1) and the effective height, z_h.

The equation for G_t was developed from the complete equations discussed in Appendix F, and assumes the resonant component of the response of the structure is negligible. The relationship for G_t as a function of total structure height is plotted in Figure 2-7 for each exposure. In this plot, the effective height of the tower is taken as 2/3 of the total height of the structure. As the resonant component is not included, Equation (2-4) yields identical values for all structure types (e.g., self-supported latticed towers, guyed towers, monopole structures, H-frame structures). This simplified equation is applicable for most transmission structure types. For taller structures, a more detailed analysis which considers the resonant component is recommended.

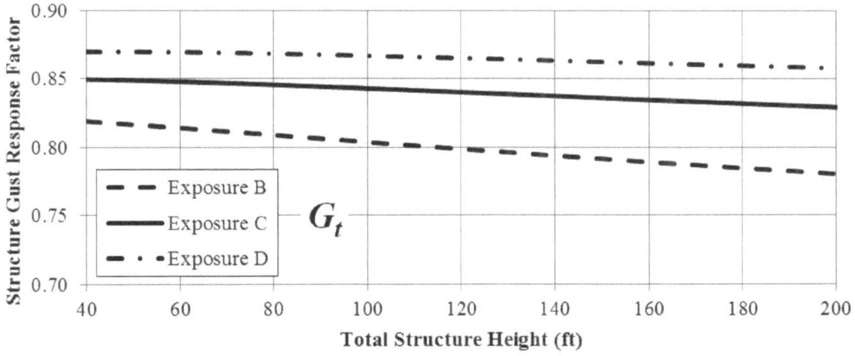

Figure 2-7. Structure gust response factor.

2.1.5.3 Gust Response Factor for Wires

The wire gust response factor, G_w, is used in Equation (2-1) for computing the peak dynamic wind loads resulting from wind on the conductors and overhead ground wires. It is given by Equation (2-5). G_w is a function of exposure category (defined in Section 2.1.4.1), design wind span between structures, and the effective height of the wires, z_k.

Equation (2-5) and the curves for G_w in Figure 2-8 were developed from the simplified equations discussed in this section and assume that the resonant response of the wires to gusting wind is negligible. The resonant component of the dynamic response is often quite small due to the lack of gust correlation in the line direction and the significance of aerodynamic damping in the wires.

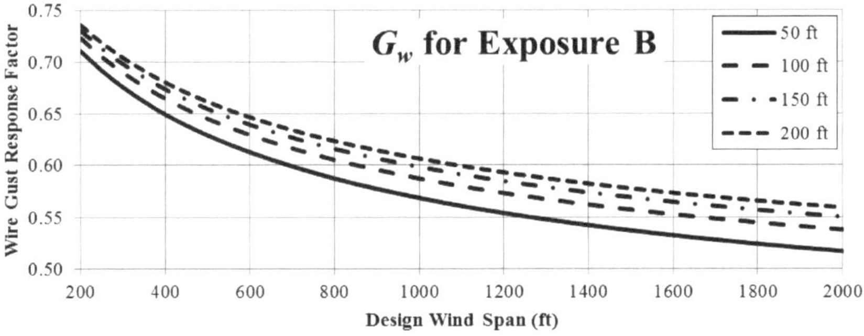

(a)

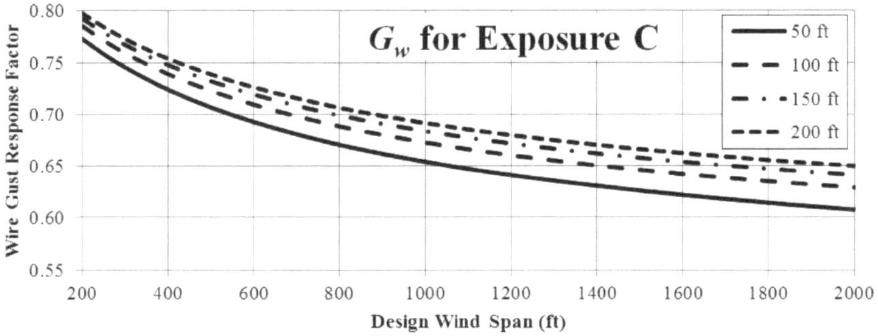

(b)

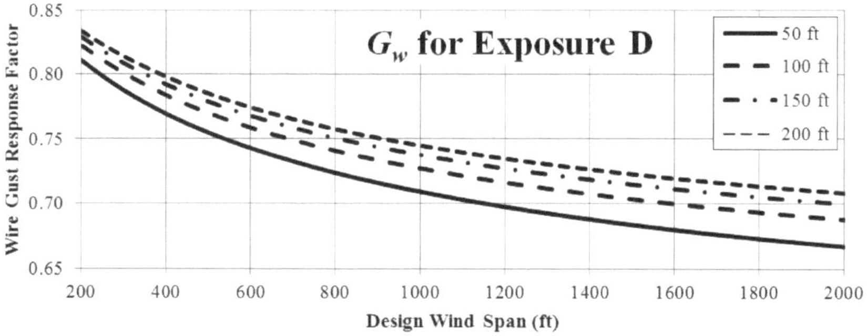

(c)

Figure 2-8. (a) Wire gust response factor for Exposure B for different effective heights, z_h; (b) wire gust response factor for Exposure C for different effective heights, z_h; (c) wire gust response factor for Exposure D for different effective heights, z_h.

2.1.6 Force Coefficient

The force coefficient, C_f, in the wind force equation [Equation (2-1)] accounts for the effects of member characteristics (e.g., shape, size, solidity, shielding, orientation with respect to the wind, surface roughness) on the resultant force. The force coefficient is the ratio of the resulting force per unit area in the direction of the wind to the applied wind pressure. It can also be referred to as the drag coefficient, pressure coefficient, or shape factor.

2.1.6.1 Factors Influencing Force Coefficients This section discusses some of the important factors in the determination of the force coefficient for a member or assembly of members. Additional theoretical background can be found in Hoerner (1958), Davenport (1960), Sachs (1978), and Mehta and Lou (1983).

2.1.6.1.1 Shape and Size Member shapes fall into two general classifications: bluff and streamlined. The forces due to wind on a bluff structure can be primarily attributed to the pressure distribution around the shape. For streamlined shapes, such as airplane wings, friction accounts for the majority of the drag force. Most buildings and engineering structures are treated as bluff bodies (MacDonald 1975, Holmes 2001).

Bluff bodies can be considered to be divided into two classes: sharp-edged and rounded. For sharp-edged members, such as rolled structural shapes, the pressure distribution around the body remains relatively constant for a given shape regardless of size or wind speed. These members are often referred to as flat-sided members. A single force coefficient is provided for flat-sided members. For rounded members, which are considered to be semi-aerodynamic, the pressure distribution around the body varies with wind speed. Above a particular wind speed, referred to as the critical wind speed, the negative pressure on the leeward side of the shape decreases in magnitude. This causes a reduction in the overall force coefficient of the member. The wind speed at which this change occurs is dependent on the Reynolds number, which is a dimensionless ratio that relates the inertia force (pressure) of the wind to its viscous force (friction). The equation for the Reynolds number is given as

$$Re = 9350\sqrt{K_z}\ V_{\mathrm{MRI}}(d_s/12) \tag{2-9}$$

where

Re = Reynolds number, referenced at standard atmosphere,
K_z = Terrain factor at height z above ground (Table 2-2),
V_{MRI} = Basic design wind speed (mph) (Section 2.1), and
d_s = Diameter of the conductor or ground wire, or the width of the structural shape normal to the wind direction (in inches).

Ice accretion on wires and structural members changes the force coefficient for these components; refer to Section 2.3.4.3, Section 2.3.5.2, and Appendix G.

2.1.6.1.2 Aspect Ratio The ratio of the length of a member to its diameter (or width) is known as the aspect ratio. Short members have lower force coefficients than long members of the same shape. The force coefficients given in Section 2.1.6.2 are applicable to members with aspect ratios greater than 40, which is typical of most transmission line structures. Correction factors for members with aspect ratios less than 40 are given in Appendix G.

2.1.6.1.3 Solidity Ratio An important parameter that influences the force coefficient for latticed truss structures is the solidity ratio of the faces. The force coefficient for the total structure is dependent on the airflow resistance of individual members and on the airflow patterns around the members. The force coefficients shown in Section 2.1.6.2 are a function of solidity ratio, Φ, which is defined as

$$\Phi = \frac{A_m}{A_o} \tag{2-10}$$

where A_m is the area of all members in the windward face of the structure (net area), and A_o is the area of the outline of the windward face of the structure (gross area).

The solidity ratio of each discrete panel in the transverse and longitudinal faces should be used for the determination of wind loads.

2.1.6.1.4 Shielding If an upstream body (e.g., structural element, component, wire) protects a downstream body from the wind, either partially or completely, it is referred to as shielding. The amount of shielding experienced by the downwind body is influenced by the solidity ratio of the upstream body, the spacing between the bodies (referred to as separation distance), and the yaw angle of the wind. The shielding factor is defined as the ratio of the force coefficient of the shielded body to the force coefficient of the unshielded body.

Shielding factors for trusses and frames are available in some codes and literature (i.e., NRCC 2010, Davenport 1960, SIA 1956). The expressions for the force coefficients of square and triangular latticed truss structures in Table 2-4 already include the effect of shielding on the downwind faces of the truss structures.

There are typically no shielding effects between individual poles of H-frames or multi-pole structures due to the low ratio of member diameter to separation distance and the small range of yaw angles for which a downwind pole would be shielded. For similar reasons, shielding effects are not typically considered for wires (e.g., bundled conductors, adjacent phases/poles), or transmission structures which are in line with one another.

2.1.6.1.5 Yawed Wind The term *yawed wind* is used to describe wind for which the angle of incidence with the face or shape is not perpendicular. The yaw angle, Ψ, is measured in the horizontal plane and is referenced as 0° for wind perpendicular to the conductors. Figure 2-9 shows an example of yawed wind and the resultant force directions. Transverse loads act perpendicular to the direction of the transmission line, while longitudinal loads act parallel to the direction of the transmission line. Expressions for yawed wind on wires and structures are provided in Section 2.1.6.2.

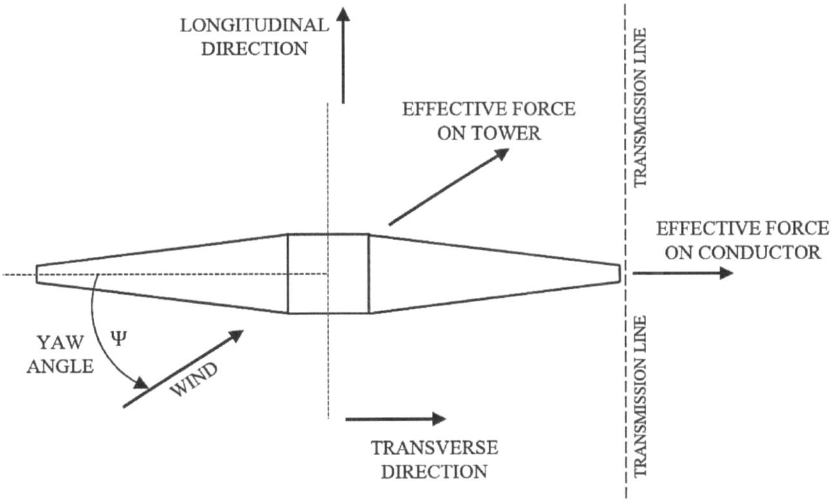

Figure 2-9. Yawed wind on a transmission line.

2.1.6.2 Recommended Force Coefficients The following sections describe the force coefficients recommended by this manual for various components of a transmission system. Other force coefficients can be used where justified by experimental data (i.e., wind tunnel testing). Additional background information on force coefficients can be found in Appendix G.

2.1.6.2.1 Conductors and Ground Wires Many designers currently use a force coefficient of 1.0 for conductors and ground wires, as indicated in NESC Rule 251 (NESC 2017). Values from wind tunnel testing, which are described in detail in Appendix G, range from 0.7 to 1.35. These data exhibit large variations in the wire force coefficient over a wide range of Reynolds numbers. Unless more definitive data based on wind force measurements are available (e.g., wind tunnel testing), a constant force coefficient value of

$$C_f = 1.0 \qquad (2\text{-}11)$$

is recommended for single or bundled conductors and for ground wires. Smaller wire sizes typically have a higher force coefficient (Appendix G). Note that if a reduced value of C_f is used on bare wires based on wind tunnel data, instances of wind loading for ice-covered wires should revert to a value of 1.0.

Equation (2-1) may be multiplied by $\cos^2\Psi$ to account for yawed wind on conductors and ground wires, in which Ψ is the yaw angle (Figure 2-9). The equation for wind force on conductors for yawed winds is given in Equation (2-12). For all yaw angles, the effective force calculated by Equation (2-12) is perpendicular to the conductor or ground wire.

$$F = QK_zK_{zt}V_{\text{MRI}}^2G_wC_fA\cos^2\Psi \qquad (2\text{-}12)$$

2.1.6.2.2 Latticed Truss Structures This manual recommends that force coefficients for square-section and triangular-section (in plan view) latticed truss structures be determined from ASCE 7-16 unless other requirements dictate the design. The relevant force coefficients are shown in Table 2-4 and Figure 2-10, as obtained from ASCE 7-16. These force coefficients account for both the windward and leeward faces, including the shielding of the leeward face by members in the windward face. Therefore, the force coefficients are multiplied by the projected area of one tower face only.

The ASCE 7-16 force coefficients for square-section and triangular-section latticed truss structures having flat-sided members are given in Table 2-4. The force coefficients given in this table for square-section structures may also be used for rectangular-section structures. For towers with round-section member shapes, the force coefficients are determined by multiplying the value calculated from Table 2-4 by Equation (2-13). Note that the correction factor for rounded members as calculated by Equation (2-13) has a limit of 1.0. The relationship of the force coefficient and solidity ratio is plotted for each cross-section type in Figure 2-10. Caution should be used when applying the force coefficients provided in Table 2-4 to sections of unique geometry, as the equations are primarily based on aerodynamic results for typical sections; additional discussion is given by Mara (2014).

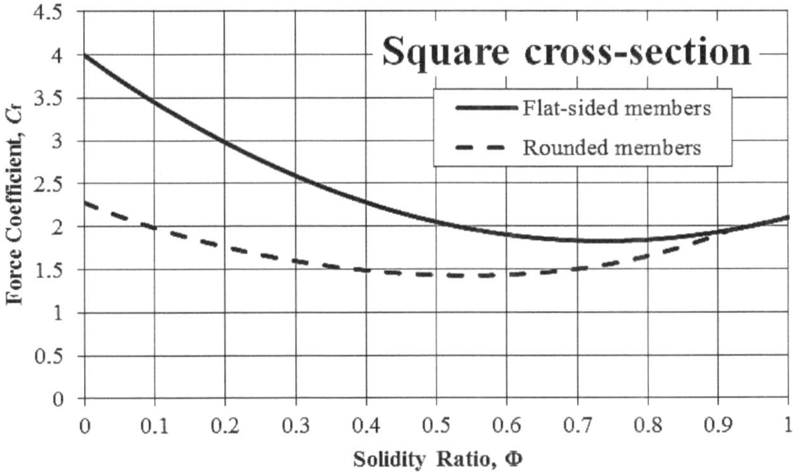

(a)

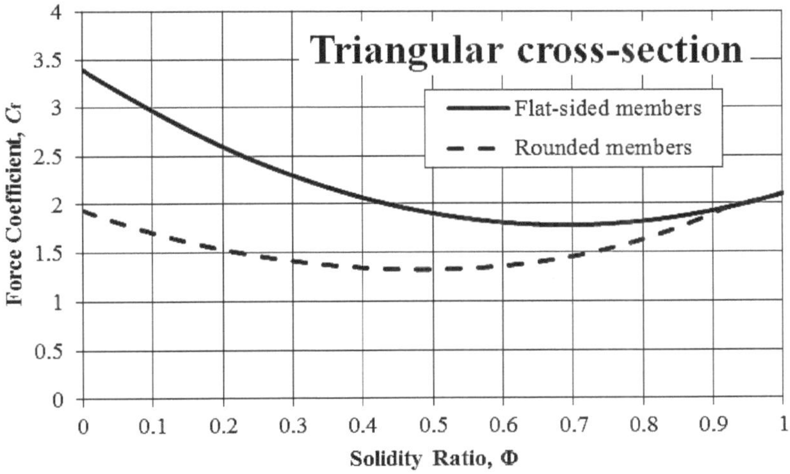

(b)

Figure 2-10. Force coefficients for (a) square; and (b) triangular cross sections.

Table 2-4. Force Coefficients, C_f, for Normal Wind on Latticed Truss Structures Having Flat-Sided Members

Tower cross-section	C_f
Square	$4.0\Phi^2 - 5.9\Phi + 4.0$
Triangular	$3.4\Phi2 - 4.7\Phi + 3.4$

Source: Adapted from ASCE 7-16 (ASCE 2017).

$$\text{Correction factor (for round members)} = 0.51\Phi2 + 0.57\text{, but not greater than } 1.0 \tag{2-13}$$

2.1.6.2.3 Latticed Truss Structures: Transverse, Longitudinal, and Yawed Wind Various methods are available for the calculation of wind loads on latticed truss structures. This manual presents two methods for the calculation of transverse, longitudinal, and yawed wind loads. The fundamentals of each method are presented in this section, along with the assumptions and key points that should be considered by the designer should either method be used. It is recognized that the two methods presented do not provide equal results; however, either method may be used by the designer provided the parameters are appropriately addressed. Either method may provide a greater wind load than the other, which is dependent on the selected parameters and application. The method resulting in the greater load does not imply correctness or that it should be considered the default method. It is the responsibility of the designer to provide justification for the method and parameters used. Other methods may be used provided good engineering judgment is applied.

The first method is based on identifying the aerodynamic properties (net area and force coefficient) of the transverse and longitudinal faces of the tower, typically in multiple sections over the height. This is commonly referred to as the "Wind on Face" method.

The second method is based on calculating the wind load on each individual member of the tower. This is commonly referred to as the "Wind on Member" method.

Method 1: Wind on Face This method starts with the segmenting of the tower geometry at a reasonable number of heights and calculating the net area and solidity ratio for each segment (or panel). The force coefficient for each panel is then calculated based on Table 2-4. The resultant wind force is then calculated for each panel following Equation (2-14a). An expression similar to Equation (2-14a) is currently used in the International Electrotechnical Commission Standard 60826 (2017).

$$F_d = QK_zK_{zt}V_{MRI}{}^2G_t\delta_\Psi(C_{ft}A_{mt}\cos^2\Psi + C_{fl}A_{ml}\sin^2\Psi) \qquad (2\text{-}14a)$$

where

F_d = Force in the direction of the wind [lb (N)],
A_{mt} = Area of all members in the face of the structure that is parallel to the line [ft^2 (m^2)],
A_{ml} = Area of all members in the face of the structure that is perpendicular to the line [ft^2 (m^2)],
C_{ft} = Force coefficient associated with the face of the structure that is parallel to the line,
C_{fl} = Force coefficient associated with the face of the structure that is perpendicular to the line,
δ_Ψ = Wind angle magnification factor equal to $[1 + 0.2\sin^2(2\Psi)]$
The other variables are as defined previously.

The resultant wind force is then decomposed into forces in the transverse and longitudinal directions, F_t and F_l, respectively, to assist with the design of each direction.

$$F_t = F_d\cos\Psi \qquad (2\text{-}14b)$$

$$F_l = F_d\sin\Psi \qquad (2\text{-}14c)$$

The presentation of this method differs from that recommended in the previous edition of this manual. While little difference exists between this method and that previously used for sections which are symmetric in plan view (i.e., square), the aerodynamics of portions of the structure which are not symmetric in plan view (i.e., cross-arm, bridge) are better captured through the current method.

As presented, the "Wind on Face" method contains some assumptions. Most notably these assumptions include

- The selection of tower face segments should represent the different structural member panel profiles for calculation of A_{mt}, A_{ml}, and the resulting solidity ratios.
- The force coefficients used (Table 2-4) are based on those available in the literature for symmetric (in plan) square and triangular sections. The design under consideration may not reflect these characteristics. However, if the extents of the face (and thus the solidity ratio) are properly selected, this method should provide acceptable values of the force coefficients.

- The equation is based on the assumption that the maximum wind load on the structure occurs for a yaw angle of 45°. This results from the wind angle magnification factor δ_Ψ in Equation (2-14a). Wind tunnel tests on models of square sections (BEAIRA 1935, Bayar 1986) and a cross-arm section (Mara et al. 2010) have demonstrated that the maximum effective wind loads are created by winds at yaw angles ranging from 30° to 45°. The magnitude of these increases will vary with tower geometry.

Method 2: Wind on Member The overall wind loads on a tower are calculated based on developing wind forces acting on each member and distributing the load among the joints at the end of each member. The wind loads on each member are then summed along the height of the tower. This method is most conveniently applied using computational techniques, as there are significantly more calculations to be performed than with the "Wind on Face" method.

This method begins with the application of wind pressure on each member based on the geometric relationship between the global wind velocity vector, the member joint-to-joint orientation with respect to the global axis, and the cross-section orientation of the member with respect to its local axis. The calculated force, F_m, is in the plane formed by the wind velocity vector (which acts in the horizontal plane) and the global member orientation. The magnitude of the wind load on each member is based on Equation (2-15)

$$F_m = QK_{z,m}K_{zt,m}V_{\mathrm{MRI}}{}^2G_tC_{f,m}A_{mo}\cos^2\alpha \tag{2-15}$$

where

F_m = Force acting perpendicular to a member, in the plane formed by the horizontal wind velocity vector and the global member orientation (lb);
$K_{z,m}$ = Velocity pressure exposure coefficient for the member, based on the average height of the member;
$K_{zt,m}$ = Topographic factor for the member, based on the average height of the member;
$C_{f,m}$ = Force coefficient of the shape of member;
A_{mo} = Maximum exposed wind area on the member, based on the length of the member and the maximum dimension exposed to wind (ft^2);
α = Incidence angle, defined as the angle formed by horizontal wind velocity vector and the plane perpendicular to the global member orientation (degrees).

The other variables are as defined previously.

The calculation of the incidence angle, α, for each member at each yawed wind direction, Ψ, requires the global member orientation to be determined from the three-dimensional geometry of the structure. The selection of the force coefficient of the member, $C_{f,m}$, should be representative of the type of member under consideration. Force coefficient values for different member shapes are provided in Appendix G.

The force acting perpendicular to the member, F_m, is then resolved into the horizontal plane based on the global member orientation. The resulting horizontal force is then subsequently resolved into the transverse and longitudinal directions for the calculation of overall wind loads. The loads resulting from wind on each member are then summed and determined at each joint along the height of the tower.

As presented, the "Wind on Member" method contains some assumptions. Most notably these assumptions include

- The selected force coefficient, $C_{f,m}$, should reflect the type of member cross-section and level of inclusion of structural and non-structural members contained in the model.
- This method assumes that the force coefficient selected for the member is consistent among all wind directions and member orientations. A single, selected value of $C_{f,m}$ will neglect the true member force coefficient as it appears to the wind for all incidence angles other than that of the profile orientation based on the respective $C_{f,m}$ selected, and therefore a conservative value should be used.
- Applying this method to members whose yaw angles are parallel to the global member orientation (i.e., horizontal members in line with the wind) will produce zero load.
- This method assumes that no shielding or flow acceleration around members occurs. These are both complex mechanisms, and the designer should be comfortable with the underlying aerodynamics if shielding effects are to be considered for a particular member or members. Additional discussion of shielding can be found in Section 2.1.6.1.4.

If the assumptions are applied, this method should provide conservative estimates of the yawed angle wind loads.

2.1.6.2.4 Pole Structures The total perpendicular or yawed wind force on single-shaft and H-frame structures is the sum of the wind forces on the individual members within the structure. Typically, transmission pole shafts and closed cross-sectional shapes exceed 1 ft in diameter, which results in a Reynolds number greater than 6×105 based on the design wind speeds specified in Figure 2-1.

Table 2-5. Force Coefficients, C_f, for Members of Pole Structures.

Member shape	C_f	Adapted from
Circular	0.9	ASCE 7-16
16-sided polygonal	0.9	James (1976)
12-sided polygonal	1.0	James (1976)
8-sided polygonal	1.4	ASCE 7-16, James (1976)
6-sided polygonal	1.4	ASCE 7-16
Square, rectangle	2.0	ASCE 7-16

Surface roughness (e.g., rough for wood, smooth for steel) will influence the force coefficients for these shapes. Attachments on pole structures, such as steps, ladders, arms, and brackets, will also influence the force coefficients. The effects of attachments and surface conditions can be significant on streamlined shapes such as circular members.

Table 2-5 lists recommended force coefficients for structural shapes commonly used for transmission pole structures. These coefficients are based on research by James (1976) and on values given in ASCE 7-16. The recommended force coefficients include the effect of typical surface roughness and attachments, such as steps, ladders, and brackets. For example, the force coefficient for a circular member is based on ASCE 7-16 assuming a rough surface. This accounts for the surface condition of a wood pole or typical steel pole attachments. For 12-sided and 16-sided polygonal shapes, the corner radius ratio term from James (1976) has been omitted to account for the effect of typical attachments.

In certain cases, it may be appropriate to select force coefficients greater than those listed in Table 2-5. Appendix G provides additional force coefficients for various shapes. The use of these or other values should be based on either design experience or research results.

2.1.6.2.5 Other Members and Components Appendix G also lists force coefficients for structural shapes based on Reynolds number, corner radius, and yaw angle. The effects of steps, ladders, arms, brackets, and other attachments are not included in the values shown in Appendix G.

2.1.7 Topographic Effects

The wind speeds which transmission line structures experience can be significantly influenced by topography. Specific recommendations on the treatment of some topographic effects are beyond the scope of this document. The designer may benefit from the advice of a meteorologist or wind engineer in situations where topographic effects are expected to be severe.

Guidelines on the effects of isolated hills and ridges on wind speeds are available (e.g., ASCE 7-16). In addition, extensive field programs and research have been devoted to the subject of boundary layer flow over hills and complex terrain (Walmsley et al. 1986, Taylor et al. 1987, Miller and Davenport 1998). Examples of topographic effects include channeling of wind, flow around mountains and hills, and flow through canyons and valleys. Note that some of these instances are treated as special wind regions in Figure 2-1.

2.1.7.1 Channeling of Wind This effect occurs where there is a natural flow of air from an unrestricted area through a restricted area, such as a mountain pass. As air is channeled into an opening or canyon, it accelerates due to the Venturi effect. This type of wind is often referred to as a local canyon wind. The wind velocity through a canyon may be as much as double that in the unrestricted areas on each side. If this condition arises along the right-of-way, the design loads should be adjusted accordingly.

Buildings may create a similar channeling effect to mountains. Generally speaking, buildings would not be a major influence on a transmission line. They could, however, alter the wind loading on one or two structures and the designer should be aware that wind channeling generated by adjacent buildings could occur in certain locations along the line.

2.1.7.2 Mountains Wind tunnel tests (Britter et al. 1981, Arya et al. 1987, Snyder and Britter 1987, Gong and Ibbetson 1989, Finnigan et al. 1990) and field experiments (Coppin et al. 1994) suggest that wind speed can increase in localized areas of mountains on the windward side as well as on the leeward side (Armitt et al. 1975). When the wind is blowing normal to a mountain ridge, the air compresses as it moves up the windward side. With any opening in the ridge, the compressed air is released and accelerates as in the case of local canyon winds.

With the appropriate combination of pressure and temperature, the wind passing over a mountain ridge accelerates on the leeward side. Accelerated winds of this type are sometimes called Santa Ana, Chinook, standing wave, or downslope winds. Several areas in the United States experience downslope winds due to their proximity to mountain ridges.

2.1.7.3 Wind Speed-up over Hills, Ridges, and Escarpments Wind speed-up due to flow over hills and escarpments is addressed in ASCE 7-16, as illustrated in Figure 2-11. The provisions apply to isolated hills or escarpments located in Exposure Categories B, C, or D. The topographic feature [two-dimensional ridge or escarpment, or three-dimensional axis-symmetric hill] is described by two parameters, H and L_h, as indicated in

Topographic Factor, K_{zt}

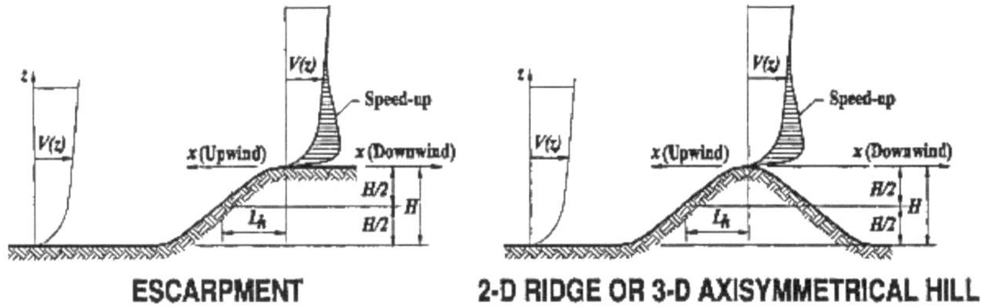

Topographic Multipliers for Exposure C

H/L_h	K_1 Multiplier			x/L_h	K_2 Multiplier		z/L_h	K_3 Multiplier		
	2-D Ridge	2-D Escarp.	3-D Axisym. Hill		2-D Escarp.	All Other Cases		2-D Ridge	2-D Escarp.	3-D Axisym. Hill
0.20	0.29	0.17	0.21	0.00	1.00	1.00	0.00	1.00	1.00	1.00
0.25	0.36	0.21	0.26	0.50	0.88	0.67	0.10	0.74	0.78	0.67
0.30	0.43	0.26	0.32	1.00	0.75	0.33	0.20	0.55	0.61	0.45
0.35	0.51	0.30	0.37	1.50	0.63	0.00	0.30	0.41	0.47	0.30
0.40	0.58	0.34	0.42	2.00	0.50	0.00	0.40	0.30	0.37	0.20
0.45	0.65	0.38	0.47	2.50	0.38	0.00	0.50	0.22	0.29	0.14
0.50	0.72	0.43	0.53	3.00	0.25	0.00	0.60	0.17	0.22	0.09
				3.50	0.13	0.00	0.70	0.12	0.17	0.06
				4.00	0.00	0.00	0.80	0.09	0.14	0.04
							0.90	0.07	0.11	0.03
							1.00	0.05	0.08	0.02
							1.50	0.01	0.02	0.00
							2.00	0.00	0.00	0.00

Notes:

1. For values of H/L_h, x/L_h and z/L_h other than those shown, linear interpolation is permitted.
2. For $H/L_h > 0.5$, assume $H/L_h = 0.5$ for evaluating K_1 and substitute 2H for L_h for evaluating K_2 and K_3.
3. Multipliers are based on the assumption that wind approaches the hill or escarpment along the direction of maximum slope.
4. Notation:
   - H: Height of hill or escarpment relative to the upwind terrain, in feet (meters).
   - L_h: Distance upwind of crest to where the difference in ground elevation is half the height of hill or escarpment, in feet (meters).
   - K_1: Factor to account for shape of topographic feature and maximum speed-up effect.
   - K_2: Factor to account for reduction in speed-up with distance upwind or downwind of crest.
   - K_3: Factor to account for reduction in speed-up with height above local terrain.
   - x: Distance (upwind or downwind) from the crest to the building site, in feet (meters).
   - z: Height above local ground level, in feet (meters).
   - μ: Horizontal attenuation factor.
   - γ: Height attenuation factor.

Figure 2-11. Topographic factor after ASCE 7-16.
Source: ASCE (2017).

Topographic Factor, K_{zt}

Equations:

$$K_{zt} = (1 + K_1 K_2 K_3)^2$$

K_1 determined from table below

$$K_2 = (1 - \frac{|x|}{\mu L_h})$$

$$K_3 = e^{-\gamma z/L_h}$$

Parameters for Speed-Up Over Hills and Escarpments						
Hill Shape	**$K_1/(H/L_h)$ Exposure**			**γ**	**μ**	
	B	**C**	**D**		**Upwind of Crest**	**Downwind of Crest**
2-dimensional ridges (or valleys with negative H in $K_1/(H/L_h)$)	1.30	1.45	1.55	3	1.5	1.5
2-dimensional escarpments	0.75	0.85	0.95	2.5	1.5	4
3-dimensional axisym. hill	0.95	1.05	1.15	4	1.5	1.5

Figure 2-11 (Continued). Topographic factor after ASCE 7-16.
Source: ASCE (2017).

Figure 2-11; H is the height of the hill, or difference in elevation between the crest and that of the upwind terrain, and L_h is the distance upwind of the crest to where the ground elevation is equal to half the height of the hill.

The topographic features may be considered in the design and location where the upwind terrain is free of such topographic features for a distance equal to $100H$ or 2 miles (3.2 km), whichever is less. The effect of wind speed-up does not need to be considered when $H/L_h < 0.2$, or when

- $H < 60$ ft (18 m) for Exposure B
- $H < 15$ ft (4.5 m) for Exposures C and D.

To account for the wind speed-up over isolated hills and escarpments that constitute abrupt changes in the general topography, a topographic factor, K_{zt}, may be applied to the transmission structures sited on the upper half of hills and ridges or near the edges of escarpments. The expression for K_{zt} is given in Equation (2-16).

$$K_{zt} = (1 + K_1K_2K_3)^2 \tag{2-16}$$

Definitions of the multipliers K_1, K_2, and K_3 are given in Figure 2-11. The multipliers are based on the assumption that the wind approaches the hill along the direction of maximum slope, causing the greatest speed-up near the crest. The value of K_{zt} should not be less than 1.0. It is not the intent of this section to address the case of wind flow over complex terrain (such as mountainous terrain, or non-isolated hills or escarpments) for which engineering judgment, expert advice, or wind tunnel testing may be required.

2.1.7.4 Canyons and Valleys Transmission lines may be subject to high winds passing through canyons, from cool air masses spilling over a ridge, or from general winds moving through a valley. Air masses spilling over into a valley can be several miles in width and may reach velocities in excess of 100 mph (160 km/h). This kind of event can occur several miles from a mountain range.

2.1.8 Application of Wind Loads to Latticed Towers

There is no standard procedure for the application of the wind forces determined from Equation (2-1) or (2-14) to the panel points of a latticed tower. Typically, the designer will follow the procedures specified by the individual utility. For example, some utilities may distribute the wind forces to the windward panel points, whereas others may distribute the wind forces to all panel points at an elevation (i.e., windward and leeward panel points). However, all utilities generally distribute wind forces to the respective member connecting joints as concentrated vector loads. A few key points should be considered when applying the calculated wind forces on different types of structures.

The wind forces determined by Equation (2-1) or (2-14), using the recommended force coefficients in this manual, account for loads accumulated by both the windward and leeward tower faces (including shielding effects). Therefore, the wind forces calculated on a single-body latticed truss system, such as a vertical double-circuit self-supported structure, can be distributed to the panel points of the structure without further consideration.

For separated latticed truss systems and individual tubular shaft members, such as a guyed portal or an H-frame structure, the windward faces of each mast should be considered as being individually exposed to the calculated wind force determined from Equation (2-1) using the appropriate force coefficients. The wind forces can then be distributed along the structure panel points according to the criteria specified by the utility. Other locations on a structure may need to be reviewed where physically separated latticed truss systems or tubular shaft members are used.

Longitudinal winds may also result in significant structural loading. This case should be considered in the design of the structure.

2.2 HIGH-INTENSITY WINDS

Tornadoes and downbursts are the high-intensity winds (HIWs) discussed in this section. HIWs are generally the result of intense, localized thermal activity that frequently accompanies a thunderstorm or squall line. These HIWs are commonly narrow-front winds with speeds greater than the sustained, broad-front, synoptic winds described in Section 2.1. HIWs do not follow the pattern and characteristics of extreme winds from which the mathematics of gust response factors in Section 2.1 were developed.

Analyses of line failures in several countries have identified HIW events as the leading cause of transmission line failures. It is possible to apply rational measures to transmission line design to increase the reliability of transmission lines impacted by the majority of HIWs in the absence of windborne debris.

2.2.1 Downbursts

2.2.1.1 Background A downburst is defined as an intense downdraft of air that induces high-velocity winds in all directions when striking the ground. Fujita (1985) defined a downburst as a mass of cold and moist air that drops suddenly from the thunderstorm cloud base, impinging on the ground surface and then transferring horizontally. The practical diameter

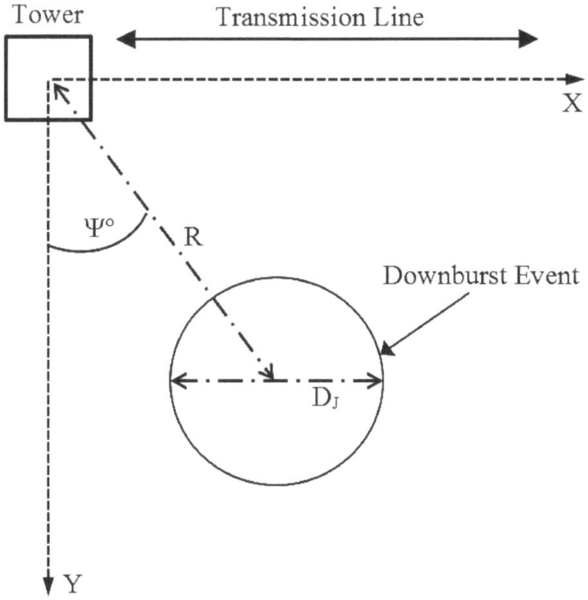

Figure 2-12. Downburst polar coordinates with respect to the tower of interest.

of this initial cold air jet can vary between 2,000 ft (600 m) and 5,600 ft (1,700 m) as indicated by Hjelmfelt (1988).

The distribution and magnitude of the forces acting on the tower and conductors depend significantly on the downburst characteristics, which are the jet velocity (V_J), the jet diameter (D_J), and the location of the downburst center relative to the tower (represented by the polar coordinates R and Ψ shown in Figure 2-12).

2.2.1.2 Downburst Wind Field The downburst outflow velocity at an arbitrary point in space has two components: a radial component V_{RD} and a vertical component V_{VR}. Using a computational fluid dynamics simulation of the spatial and time variations of the wind field associated with downbursts (Hangan et al. 2003, Shehata and El Damatty 2007) concluded that the vertical velocity component has a negligible effect on the structural response in comparison to the radial velocity component. As such, the current section focuses on describing the wind field associated with the radial velocity. Figure 2-13 illustrates the variation with height of the radial velocity, normalized with respect to the jet velocity (El Damatty and Elawady 2015). The maximum velocity occurs at an R/D_J value of approximately 1.3. The velocity profile in Figure 2-13 is provided in terms

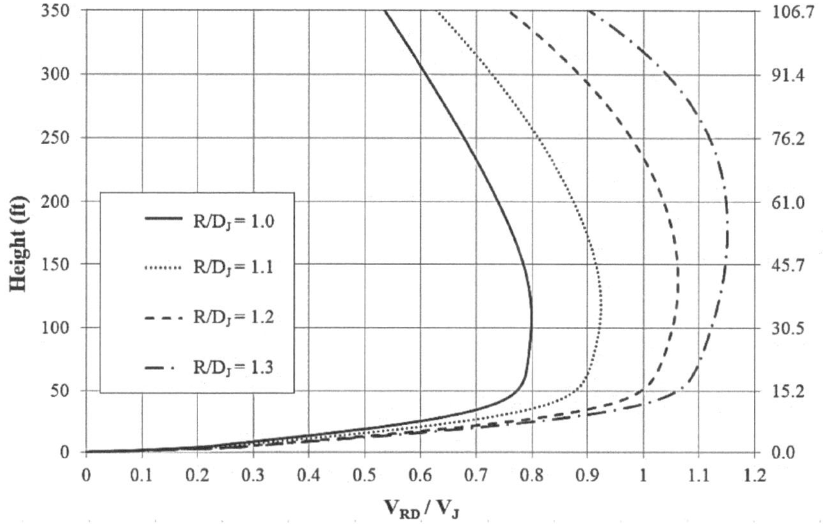

Figure 2-13. Vertical profile of the radial outflow wind associated with a downburst considering D_J *= 1,640 ft (500 m).*

of downburst jet velocity, V_J, which represents the speed at which the downburst event impinges on the surface of the earth. In view of the information found in the literature (e.g., Fujita 1985, Orwig and Schroeder 2007, CIGRÉ 2009), a value for V_J of 112 to 157 mph (50 to 70 m/s) is recommended.

The span reduction factor commonly used to adjust the loads applied on the conductor spans is closer to unity when compared with that in synoptic winds (Holmes et al. 2008, Aboshosha and El Damatty 2013). One approach for simulating critical downburst load cases for transmission line structures is provided in Appendix K.

2.2.2 Tornadoes

2.2.2.1 Background The usual perception of a tornado is of an overwhelming event destroying all in its path and defying resistance. Although most tornadoes are capable of causing severe damage to houses, mobile homes, and automobiles, most engineered structures in the vicinity of tornadoes survive without major damage. Transmission line structures can be designed with sufficient strength to resist wind loads incurred in a majority of tornado events (F0–F2). However, designing for severe tornadoes (i.e., F3 to F5) may be prohibitive due to significantly higher cost. For these severe tornadoes, the focus of the designer changes from resisting the HIWs to failure containment.

Tornadoes occur on most subtropical and temperate landmasses around the world. On average, 800 to 1,000 tornadoes occur each year in the contiguous United States, and the activity zone extends well into Canada. The total number of reported tornadoes in 1° squares of latitude and longitude for a 30-year period (1950 to 1980) is shown in Figure 2-14. A 1° square contains about 4,000 mi^2 (10,000 km^2). More recent information on tornado characteristics, occurrence, and forecasting can be obtained at the online Storm Prediction Center, which is maintained by the National Oceanic and Atmospheric Administration (NOAA)/National Weather Service.

Fujita and Pearson (1973) have developed a rating (F0 to F5) to categorize tornadoes by their intensity and size. This method assigns a numerical value of the F-scale to each tornado based on the appearance and extent of damage. The F-Scale and the associated path length, path width, and wind speed are shown in Table 2-6. The wind speed provided the fastest quarter-mile speed, assumed at 16 to 33 ft (5 to 10 m) above the ground level, and also as 3-second gust wind speed. Also provided are the number of tornadoes, percentage, and cumulative percentage for each scale. It is noted that 86% of the tornadoes are assigned to the scale of F2 or smaller; the F2 rating corresponding to a 3-second gust wind speed of approximately 160 mph (70 m/s) or less.

The probability of a tornado strike at a given point is very small (McDonald 1983), even in areas of tornado prevalence. However, the probability of a transmission line being crossed by a tornado is significant (Twisdale

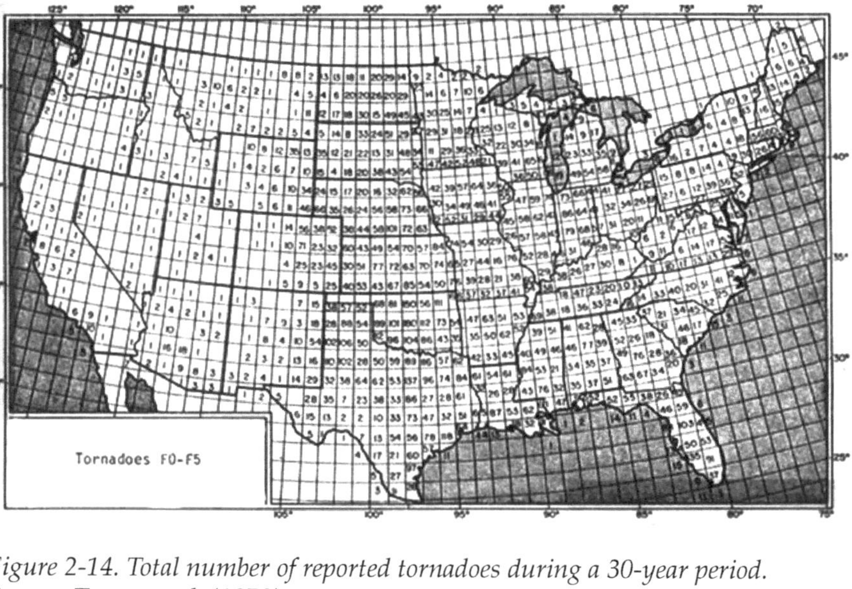

Figure 2-14. Total number of reported tornadoes during a 30-year period. Source: Tecson et al. (1979).

Table 2-6. Tornado Wind Speed, Path Length, Path Width, and Frequencies for F-Scale in the United States

Scale	Fujita Scale						
	Fastest quarter-mile	3-second gust	Path length	Path width	Tornado frequencies (1916–1978)*		
	Wind speed	Wind speed			Number of tornadoes	Percentage	Cumulative percentage
F0	72 (mph) 32.2 (m/sec)	45–78 (mph)	<1.0 (mi) 1.61 (km)	<50 (ft) 15.2 (m)	5,718	22.9	22.9
F1	73–112 (mph) 32.6–50 (m/s)	79–117 (mph)	1.1–3.1 (mi) 1.8–5.0 (km)	51–170 (ft) 15.5–52 (m)	8,645	34.7	57.6
F2	113–157 (mph) 50.5–70.2 (m/s)	118–161 (mph)	3.2–9.9 (mi) 5.1–15.9 (km)	171–530 (ft) 52.1–161.5 (m)	7,102	28.5	86.1
F3	158–206 (mph) 70.6–92.1 (m/s)	162–209 (mph)	10–31 (mi) 16–50 (km)	531–1,670 (ft) 1,61.8–509 (m)	2,665	10.7	96.8
F4	207–260 (mph) 92.5–116.2 (m/s)	210–261 (mph)	32–99 (mi) 51–159 (km)	1,671–4,750 (ft) 509.3–1,447 (m)	673	2.7	99.5
F5	261–318 (mph) 116.7–142.2 (m/s)	262–317 (mph)	100–315 (mi) 160–507 (km)	4,751–6,000 (ft) 1,448–1,829 (m)	127	0.5	100

**Source:* Tecson et al. (1979)

1982). The fact that the width of the path is very narrow for most tornadoes, however, makes it possible to improve transmission line resistance to the majority of tornadoes. Almost all tornadoes can engulf a house or small structure, but very few have a width that will load the full span of a transmission line, with the exception of spans less than perhaps 500 to 600 ft (150 to 200 m).

2.2.2.2 Tornado Wind Field The wind pattern within a tornado is composed of circular winds combined with translation, the highest velocities being where the rotary and translation components add together (Abbey 1976; Mehta et al. 1976; Minor et al. 1977; Wen and Chu 1973; Hangan and Kim 2008; Hamada et al. 2010; Hamada and El Damatty 2011, 2015). An idealized pattern of tornado wind velocities is shown in Figure 2-15. The tornado wind velocity at a certain point in space has three components: tangential velocity, radial velocity, and uplift (axial) velocity.

Field data for the 1998 Spencer, South Dakota, F4 tornado and for the 1999 Mulhall, Oklahoma, F4 tornado were used to validate the numerical simulations of F4 and F2 tornadoes (Hangan and Kim 2008, Hamada et al. 2010) that are discussed subsequently and in Appendix K.

Idealized vertical profiles for the axisymmetric average of tangential, radial, and vertical velocity components below 330 ft (100 m) are provided in Figures 2-16 to 2-18 at three radial distances, r, where r is the distance

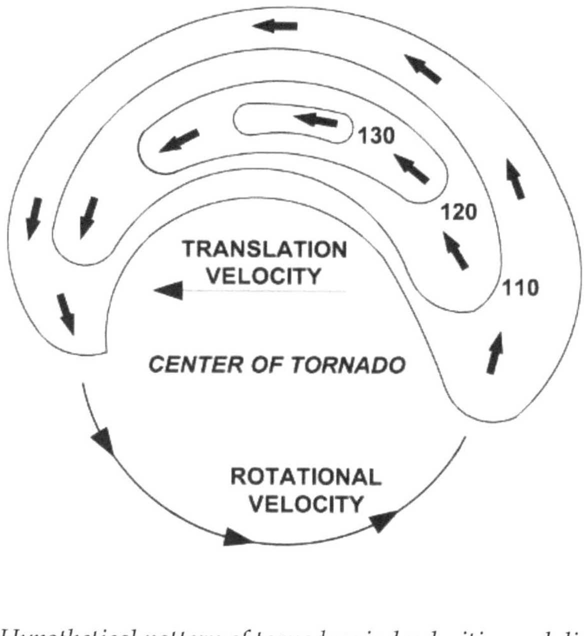

Figure 2-15. Hypothetical pattern of tornado wind velocities and directions.

measured from the center of a tornado. The radial velocity changes direction with height, where negative values act in an inward direction and positive values act in an outward direction. A positive value for the axial or vertical component indicates an upward velocity. One approach for simulating critical tornado load cases for transmission line structures is provided in Appendix K.

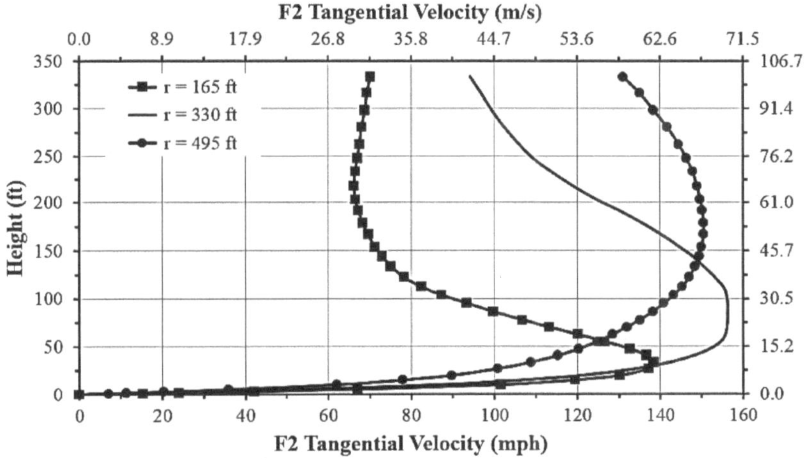

Figure 2-16. Idealized vertical profiles of tangential velocity component for three radial distances from F2 tornado center.

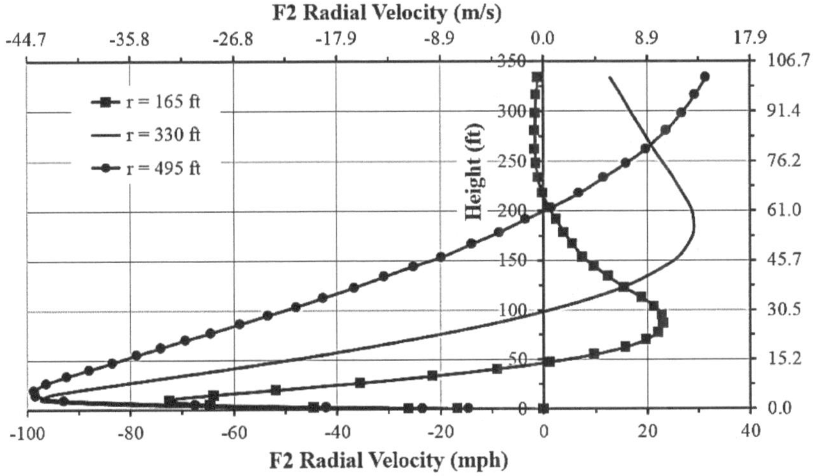

Figure 2-17. Idealized vertical profiles of radial velocity component for three radial distances from F2 tornado center.

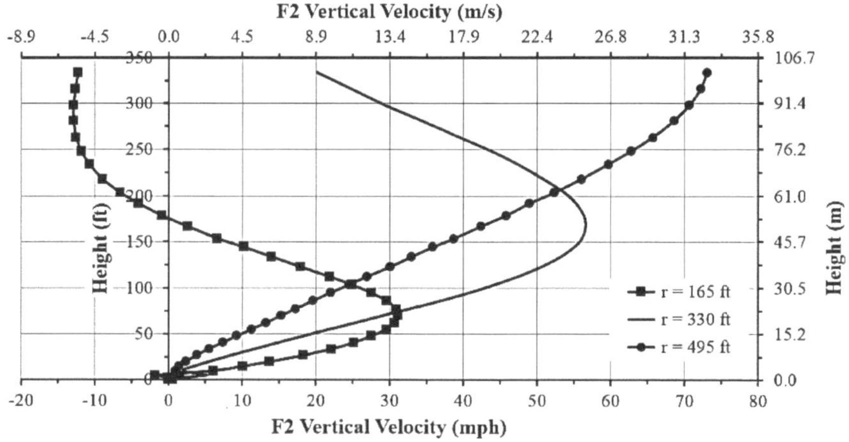

Figure 2-18. Idealized vertical profiles of vertical velocity component for three radial distances from F2 tornado center.

2.3 ICE AND WIND LOADING

2.3.1 Introduction

Ice accretion on a transmission line is often a governing loading criterion in structure design. In addition to imposing substantial vertical loads on the structural system, ice buildup on the wires and conductors presents a greater projected area exposed to the wind and affects the force coefficient. The resultant load on the wires and conductors may cause significantly greater wire tensions compared to the bare conductor conditions. Meteorological data suggest, and a survey of utility practice (ASCE 1982) confirms, that it is prudent to include concurrent ice and wind loadings in the load criteria of transmission structure designs throughout most of the United States.

The following discussion provides general guidance for the selection of ice with concurrent wind loads. Where more detailed icing data have been compiled for a service area, those data should take precedence over the information in this manual. Electric utilities are urged to develop concurrent ice and wind loading criteria specifically for their service regions based on historical data.

2.3.2 Categories of Icing

Ice can be classified by either its method of formation or its physical characteristics. Precipitation icing from freezing rain or freezing drizzle is the most common icing mechanism.

The inglaze ice that forms in these conditions is usually clear but may also be translucent because of included air bubbles.

In-cloud icing is caused by supercooled cloud droplets, carried by the wind, colliding with a surface. The ice that forms ranges from hard, clear glaze to softer, lower-density white rime ice containing entrapped air. In-cloud icing may occur in regions with level terrain, but is more frequently associated with mountainous areas, occurring on both exposed summits and upslopes.

Snow, both wet and dry, may adhere to wires by capillary forces, freezing and sintering, forming a cylindrical sleeve around the wire. The density of accreted snow depends on the wind speed and wetness of the snow.

Hoarfrost is an accumulation of ice crystals formed by the direct deposition of water vapor from the air onto a structure. The amount of ice accreted by vapor deposition does not impose significant loads on structures.

It is important that the transmission line engineer be aware of the icing conditions (i.e., freezing precipitation, in-cloud icing, or sticky snow) that may occur along the route of a proposed transmission line. Ice accretions produced by freezing rain rarely exceed a thickness of a few inches, whereas lower-density accretions due to in-cloud icing and sticky snow can build to thicknesses of a foot or greater. Furthermore, in-cloud icing can produce significant unbalanced loadings between adjacent spans with different wind exposures. The designer would benefit from the advice of a meteorologist in regions where in-cloud icing may be severe.

Section H.1 in Appendix H provides additional information on the meteorological conditions that are associated with the various types of icing and properties of the ice accretions.

2.3.3 Design Assumptions for Ice Loading

The four categories of icing (glaze, in-cloud, snow, and hoarfrost) cover the spectrum of the icing phenomenon. The distinctions made by definition of each category may not be identifiable in practice. There can be an overlap of more than one type of icing condition, such as snow and freezing rain, or in-cloud icing and freezing drizzle. In specifying ice loadings, the accretion density should be noted and is typically assumed to be uniform with thickness.

For simplicity, the design ice thickness is specified as an equivalent uniform radial thickness over the length of the wire. However, natural ice accretions may be uniform, elliptical, crescent-shaped, pennant-shaped, or have attached icicles.

2.3.4 Ice Accretion on Wires Due to Freezing Rain

2.3.4.1 Historical Ice Data As weather stations do not collect data on ice thickness, ice accretion models based on meteorological data are often used to estimate ice thickness. Where modeled or measured ice thickness and concurrent wind speed data are available, the calculated nominal ice thickness is defined as t (which is t_{100} or t_{MRI}). The concurrent wind speed is used to determine the concurrent transverse wind load that is combined with the vertical load due to the weight of the ice.

2.3.4.2 Using Ice Accretion Maps In areas where local historical icing data are not available, the glaze ice maps given in Figures 2-19 through 2-23 can be used with some limitations. These maps show 100-year MRI ice thicknesses due to freezing precipitation with concurrent 3-second gust wind speeds V_3-sec at 33 ft (10 m) above ground for the continental United States and Alaska. These maps are revised from the maps in the previous edition of this manual (ASCE 2010a) using data from weather stations in Canada near the US border and using a 100-year MRI instead of a 50-year MRI. The stations used to map Alaska are shown in Figure 2-23, and stations used to map the lower 48 states are shown in Appendix H, Figure H-1. The glaze ice thicknesses shown in these figures do not include in-cloud icing or sticky snow accretions, which are caused by meteorological conditions that may produce significantly different loading patterns (see Appendix H, Section H.5).

Additional maps showing the 50-year and 300-year MRI ice thicknesses due to freezing precipitation with concurrent 3-second gust wind speeds are provided in Appendix L. The mapped ice thicknesses and wind speeds are based on Exposure C, but should also be used for Exposures B and D.

The amount of ice that accretes on a wire depends on the wind speed at the wire height. Design thicknesses of glaze ice t_z for heights z above ground can be obtained from Equations (2-17a) and (2-17b).

$$t_z = t_{MRI}\left(\frac{z}{33}\right)^{0.10} \quad \text{for } 0 \text{ ft} < z < 900 \text{ ft} \tag{2-17a}$$

$$t_z = t_{MRI}\left(\frac{z}{10}\right)^{0.10} \quad \text{for } 0 \text{ m} < z < 275 \text{ m} \tag{2-17b}$$

where

t_{MRI} = Nominal ice thickness (e.g., t_{100} for 100-year MRI),
t_z = Design ice thickness at height z, and
z = Height above ground (ft in customary units [Equation (2-17a)]; m in metric units [Equation (2-17b)]).

The exponent in Equation (2-17) is based on a power law increase of wind speed with height for Exposure Category C (open country terrain).

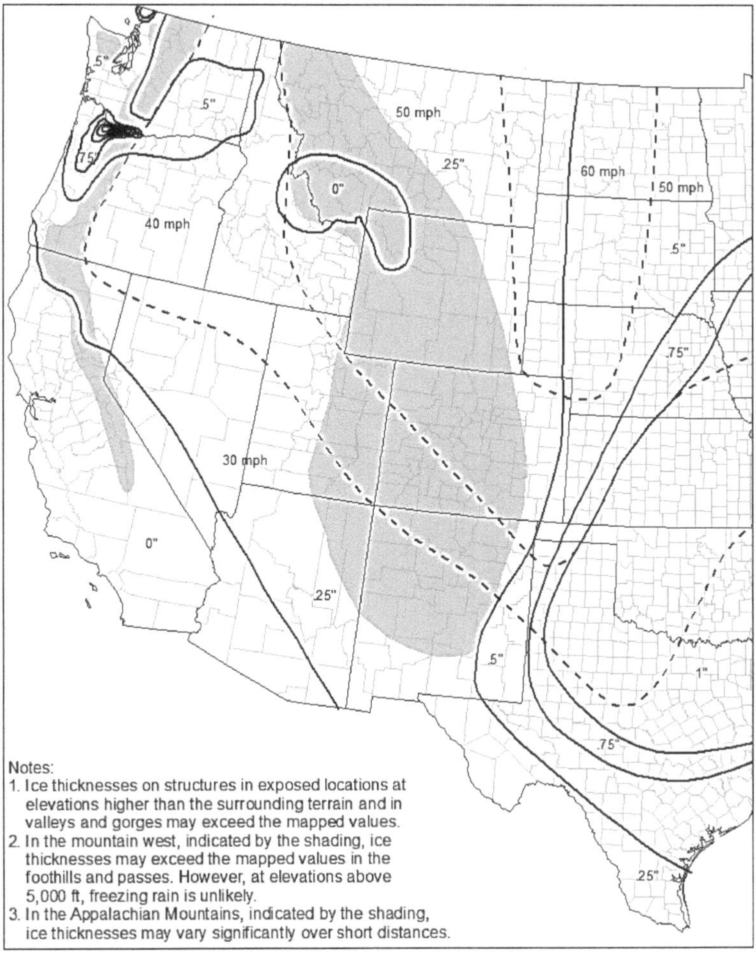

(a)

Figure 2-19. 100-year MRI radial ice thickness (in.) from freezing rain with concurrent gust wind speeds (mph) at 33 ft (10 m) aboveground in Exposure C for the (a) western; and (b) eastern United States.

At sites that tend to be windy or where the wind speed increases rapidly with height, the ice thickness gradient will be more pronounced than indicated by Equation (2-17). The concurrent gust wind pressure is also increased with height above ground using Equation (2-3) and the power law exponents indicated in Table 2-1. Ice thicknesses on a ridge, hill, or

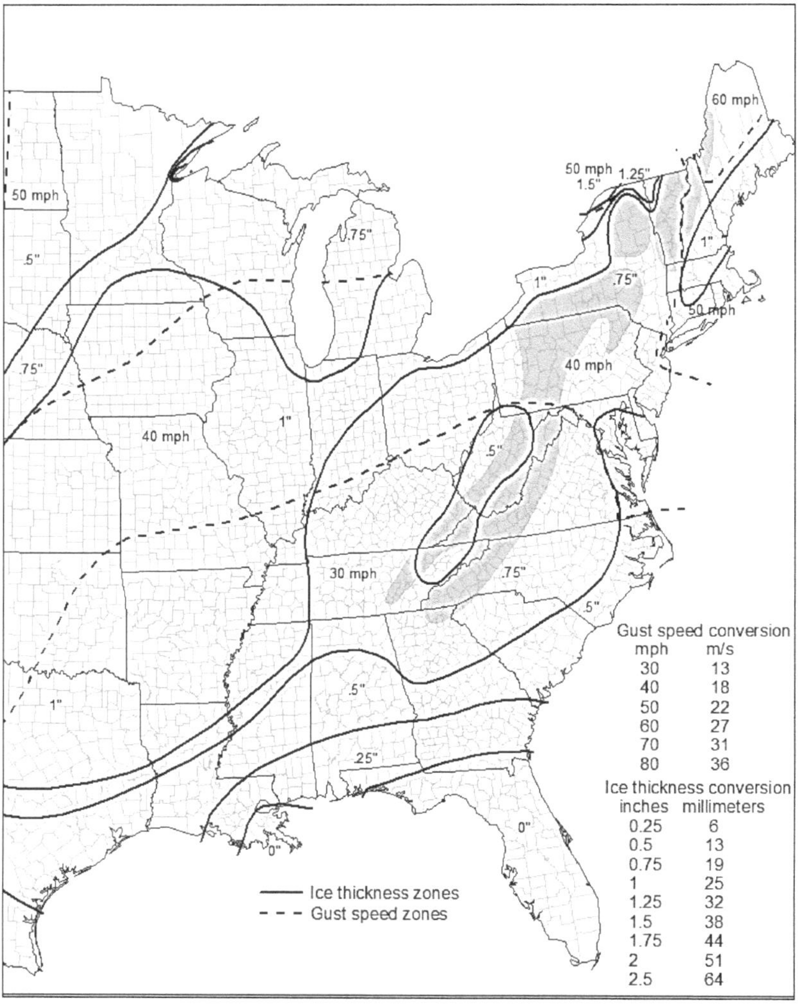

(b)

Figure 2-19. (Continued) 100-year MRI radial ice thickness (in.) from freezing rain with concurrent gust wind speeds (mph) at 33 ft (10 m) aboveground in Exposure C for the (a) western; and (b) eastern United States.

escarpment will be greater than those in level terrain due to wind speed-up effects. The topographic factor for the ice thickness on isolated ridges, hills, or escarpments is $K_{zt}0.35$, where K_{zt} is obtained from Equation (2-16). It should be noted that as ice thickness and concurrent wind are affected by height above ground and topography, there are additional uncertainties involved with this process than suggested by the preceding calculations.

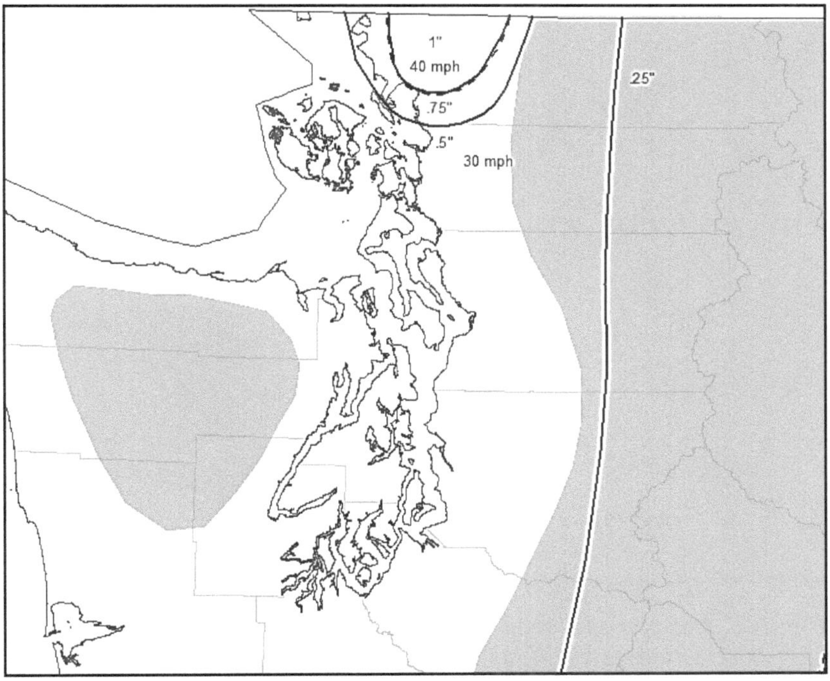

Figure 2-20. 100-year MRI radial ice thickness (in.) from freezing rain with concurrent gust wind speeds (mph) at 33 ft (10 m) aboveground in Exposure C: Puget Sound detail.

For areas not covered by Figures 2-19 though 2-23 and areas where in-cloud icing or sticky snow is the most severe icing mechanism, other sources of information must be consulted to determine design ice thicknesses; refer to Appendix H, Sections H.4 and H.5, for additional information. Figures 2-19 through 2-23 represent glaze ice thickness values at single points, and do not include spatial effects (refer to Appendix B).

In areas where little information on ice loads is available, it is recommended that a meteorologist familiar with atmospheric icing be consulted. Factors to be kept in mind include that taller structures may accrete more ice due to higher winds and colder temperatures aloft, and that influences of elevation, complex relief, proximity to water, and potential for unbalanced loading are significant.

2.3.4.3 Ice with Concurrent Wind Loading The ice thicknesses due to freezing rain in Sections 2.3.4.1 and 2.3.4.2 are uniform radial glaze ice

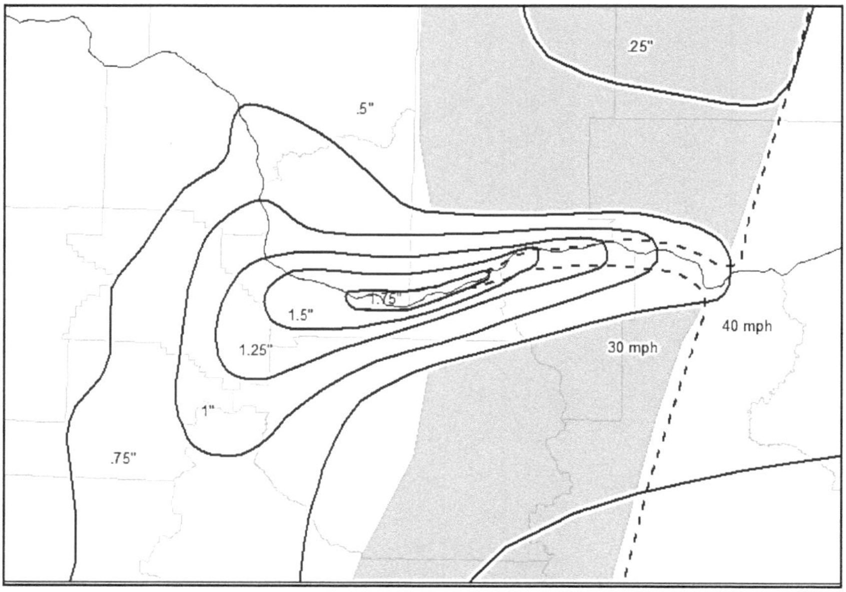

Figure 2-21. 100-year MRI radial ice thickness (in.) from freezing rain with concurrent gust wind speeds (mph) at 33 ft (10 m) aboveground in Exposure C: Columbia River Gorge detail.

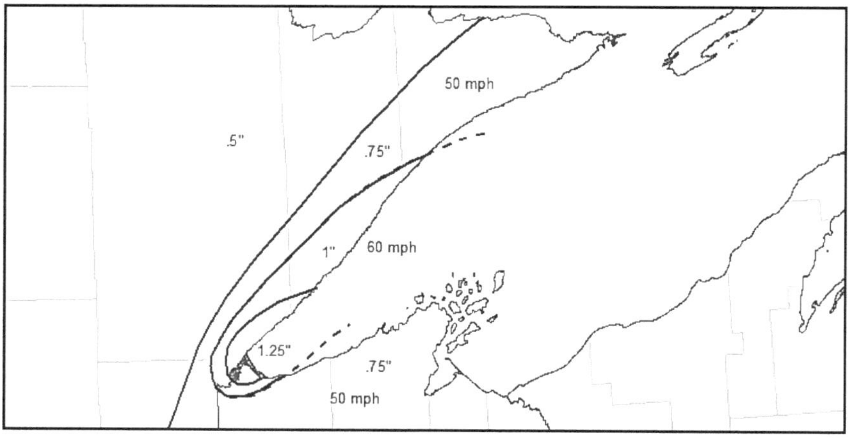

Figure 2-22. 100-year MRI radial ice thickness (in.) from freezing rain with concurrent gust wind speeds (mph) at 33 ft (10 m) aboveground in Exposure C: Lake Superior detail.

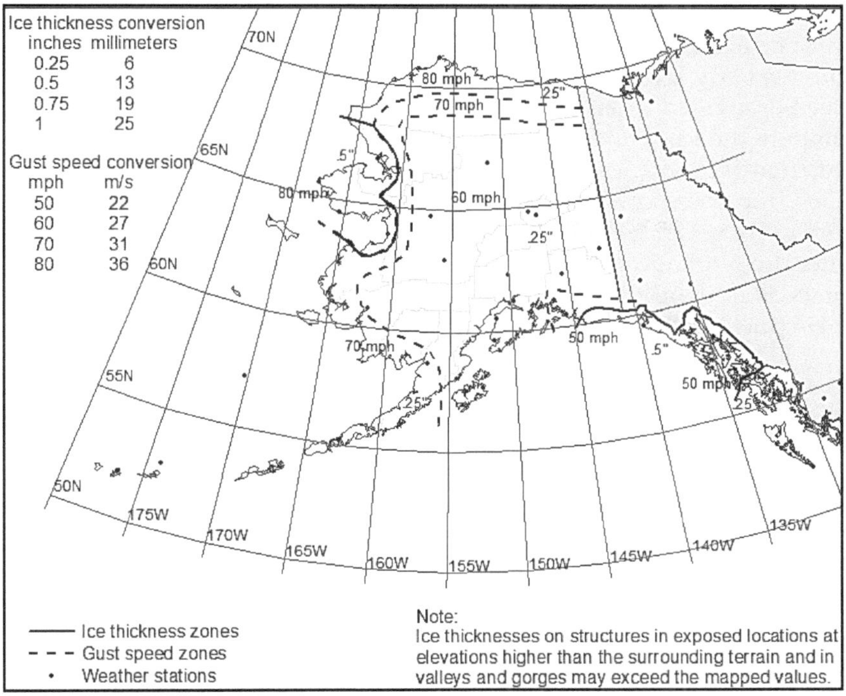

Figure 2-23. 100-year MRI radial ice thickness (in.) from freezing rain with concurrent gust wind speeds (mph) at 33 ft (10 m) aboveground in Exposure C: Alaska.

thicknesses. Using an ice density ρ_i = 56 pcf for bubble-free glaze ice, the linear ice load on a wire is calculated as

$$W_i = Q_i(d + t_z)t_z \tag{2-18}$$

where

W_i = Weight of glaze ice (lb/ft customary units, N/m metric units).
Q_i = Constant to convert ice thickness to weight (1.2435 customary units, 0.0282 metric units),
d = Diameter of bare wire (in. customary units, mm metric units), and
t_z = Design ice thickness at height z above ground (in. customary units, mm metric units)

Ice accretion on a wire can substantially increase its projected area. The transverse load due to wind acting on ice-covered wires acts concurrently with the vertical load due to the weight of the ice. The 3-second gust wind speeds provided in Figures 2-19 through 2-23 should be used with t_z to compute the ice with concurrent wind load using the methodology presented in Section 2.1. When calculating forces due to wind on ice-covered wires, the force coefficient is dependent on the shape of ice buildup (McComber et al. 1982). However, typical force coefficients of ice-covered wires are not known. Some organizations recommend using force coefficients other than 1.0 for wires covered with ice (IEC 2003b, ISO Standard 12494 (ISO 1999)).

2.3.4.4 Design Temperatures for Freezing Rain Temperature is an important consideration in calculating the tension of wires. The design temperatures concurrent with the design ice with concurrent wind loads due to freezing rain are mapped in Figures 2-24 and 2-25. These temperatures are for ice thicknesses for all MRIs. It is recommended to use

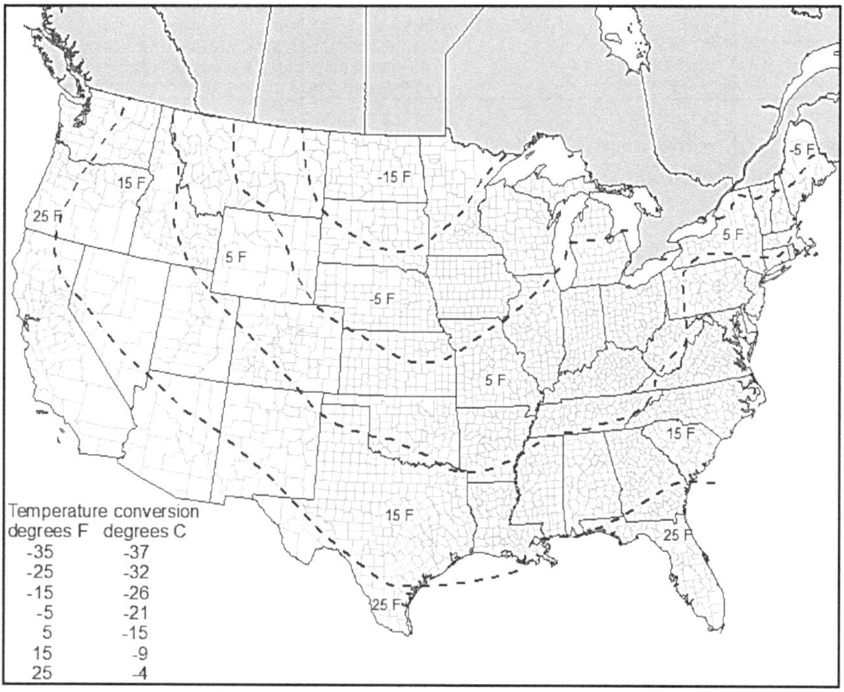

Figure 2-24. Temperatures concurrent with ice thickness attributable to freezing rain: 48 contiguous states.

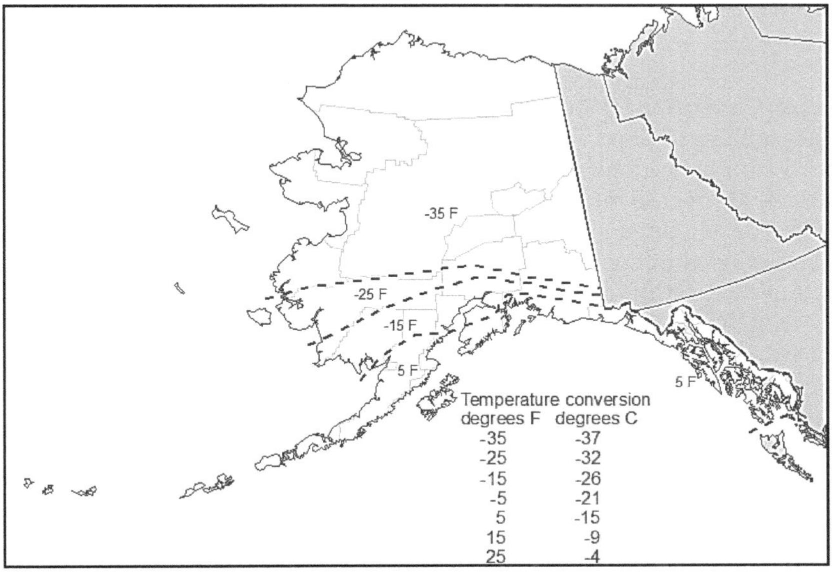

Figure 2-25. Temperatures concurrent with ice thickness attributable to freezing rain: Alaska.

either these values or 32 °F (0 °C), whichever results in the maximum load effect. A temperature of 32 °F (0 °C) should be used in Hawaii.

The temperatures shown on these maps were determined by tracking the minimum temperature that occurred with the modeled maximum ice load in each freezing rain event. The sample of minimum temperatures for all the events used in the extreme value analysis of ice thickness was then analyzed to determine the 10th percentile temperature (i.e., the minimum temperature that is exceeded by 90% of the recorded minimum temperatures). These 10th percentile temperatures are shown on the maps. In areas where the temperature contours were near the wind or ice thickness contours, they were moved to coincide with, first, the concurrent wind boundaries, and second, the ice zone boundaries.

2.3.5 Ice Accretion on Structural Members

2.3.5.1 Vertical Loads Ice accretion on the structural members themselves is typically not included in design. For the design of bracing members of latticed structures and cross-arms, the construction and maintenance loads recommended in Chapter 3, Section 3.1 will generally impose design stresses greater than the bending stresses resulting from the vertical weight of ice-covered members. For vertical supports (e.g., pole shaft or leg angle), the additional axial load due to ice on the member does not significantly add to the member stress.

2.3.5.2 Concurrent Wind Loads Ice accretion on the structure may increase the projected area of the structure exposed to wind loading. For broad-profile structural members (e.g., pole sections), the fractional increase in overall projected area due to ice from precipitation icing is small. For angle members, the increased area may be partially offset by a reduction in the force coefficient due to the streamlining effect of the ice coating on the bluff angle member. Thus, for transmission line structures, it is usually not necessary to design for the increase in the projected area of a structure due to ice buildup on its members from precipitation icing.

2.3.6 Unbalanced Ice Loads

Although the principal design loading combination is for the same ice thickness and wind speed applied to all spans, unequal ice loading and wind speed should also be considered in design. Ice thicknesses and concurrent wind speeds may differ from one span to the next, typically when the exposure of a transmission line changes as it traverses a hill or ridge. Generally, tangent structures with longer suspension insulator strings will not experience significant longitudinal conductor loads due to unbalanced ice loads; however, suspension structures with short insulator strings, and in particular shield wire attachments with short hardware assemblies, may transfer most of the imbalance to the structure. The designer is referred to Section 3.1 and Appendix I of the manual.

2.3.7 Ice Accretion on Aerial Marker Balls or Similar Devices

Ice accretion due to freezing rain on aerial marker balls should be taken into account during design. The weight of the ice increases the tension in the wires and the vertical load on their supporting structures. The ice density, ρi, is the same as that applied to the wire. The marker ball ice load is determined using the weight of glaze ice formed on the projected surface of the aerial marker. Since the ice thickness specified for accretion on wires is calculated using the diameter to perimeter ratio of $1/\pi$ to convert ice accretion on a flat surface to a cylindrical surface, the wire ice thickness is multiplied by π as indicated in Equation (2-20). The weight of ice on an aerial marker ball is calculated using Equations (2-19) through (2-21)

$$W_i = V_i \rho_i \tag{2-19}$$

$$V_i = \pi t_z A_s \tag{2-20}$$

$$A_s = \pi r^2 \tag{2-21}$$

where

A_s = Projected area of aerial marker ball [Equation (2-21)],
V_i = Volume of ice accreted on the aerial marker ball [Equation (2-20)],
t_z = Design wire ice thickness,
r = Radius of the aerial marker ball, and
ρ_i = Ice density.

CHAPTER 3
ADDITIONAL LOAD CONSIDERATIONS

3.0 INTRODUCTION

Transmission line designers should consider loads and potential failure scenarios from sources other than the weather-related events described in Chapter 2. This chapter addresses additional structure and wire system loads resulting from unbalanced wind and ice, broken conductors, construction and maintenance, and other causes. These loads apply to both new installations and modifications to existing facilities. Site-specific conditions such as landslides, ice flow, frost heave, and flooding are not addressed but should be considered during design.

3.1 LONGITUDINAL LOADS, LINE SECURITY, AND FAILURE CONTAINMENT

3.1.1 Longitudinal Loads

In addition to the transverse and vertical climatic loads discussed in Chapter 2, transmission line structures are subjected to longitudinal loading due to an imbalance in wire tension or a failed component. These imbalances can originate from a number of sources and may appear when the wire system is intact or when it has been compromised. When not addressed in the design of the structure and line, unabated longitudinal loading can lead to catastrophic or cascading failures.

3.1.2 Unbalanced Loads on Intact Systems

On suspension structures, the tension in the wire in adjacent spans can differ due to variable wind speeds or unequal ice accumulation. Wind can vary from one geographical point to the next due to changes in terrain and elevation, particularly on long spans or in mountainous regions. Unequal accumulation of ice can also occur for the same reasons as observed by in-cloud icing. Lines with limited ability to transfer slack, such as those with shorter suspension insulators, spans with little sag, and binding of hardware, can limit the ability of the suspension insulators to swing sufficiently to balance tensions within the wire system.

Strain and dead-end structures are used when a change in tension is warranted due to line design requirements. On these structures, conductors are terminated on horizontal insulator strings, which apply tension directly into the structure. Strain and dead-end structures must be designed to resist these tension imbalances.

Ground wires are typically assumed to be rigidly attached to all structures, so an unbalanced tension at this location should be considered for all structures.

3.1.3 Longitudinal Loads due to Non-Intact Wire Systems

A component failure, such as a wire or insulator, or structure collapse can create severe load imbalances in the wire system capable of causing the partial or complete failure of the adjacent structures.

3.1.4 Failure Containment and Line Security Loads

Failures due to unbalanced loading and a broken wire system may continue to propagate along the line for a significant distance, resulting in a catastrophic failure. These cascading failures of transmission lines cause significant damage and large economic losses because they may destroy complete sections of a line, requiring weeks or months of repair (Ostendorp 1997). One method to prevent these cascading failures is to include line security loads in the design of transmission structures to provide longitudinal strength. This can be accomplished by including an unbalanced ice case and/or a broken wire load (BWL) case. Typically, only one phase wire or the shield wire is considered to be broken for each line circuit.

Many structure types, such as rigid square-based latticed towers, longitudinally guyed V, Y, delta, and portal structures, and single-pole structures, can be designed with increased longitudinal strength for a relatively small increase in initial cost. The criticality of the line and the detrimental effects on the electric system should be considered when determining the magnitude of the load or whether to apply a line security load at all. It may be prudent to design structures to minimize cascades and help reduce restoration time and cost.

Designers of new lines should be aware of the consequences of a single structure failure and a potential cascade. In addition to or instead of incorporating line security loads or when structures with limited longitudinal strength are used, localized structure hardening through the use of "stop" or "anti-cascading" structures may be an effective and economical way to minimize the impact of cascades. Failure containment philosophies differ by region, transmission line owner, and system operator. The length and importance of the line, longitudinal strength of the suspension structures, terrain, restoration time, emergency stocking levels, cost, right-of-way access, and proximity to highways and railways will influence the length of containment sections. Recommended distances between failure containment structures vary, but separations between 2 and 10 miles (3.2 and 16.1 km) are common.

Several accepted analytical methods can be used to estimate longitudinal design loads to prevent cascading failure. These include the Residual Static Load (RSL) Method, the Bonneville Power Administration (BPA) method, Broken Wire Load (BWL) method, and the EPRI method. Refer to Appendix I for a discussion of these methods and other failure containment design considerations.

3.2 CONSTRUCTION AND MAINTENANCE LOADS

3.2.1 General

Construction and maintenance (C&M) loads are directly related to work methods. Construction loads are imposed on structures during assembly and erection, and from installation of ground wires, insulators, conductors, and line hardware. Maintenance loads are those loads applied to the structures resulting from scheduled or emergency inspection and/or replacement of all or part of a structure, ground wire, insulator, conductor, or conductor hardware system. Knowledge of construction and maintenance work methods is required to develop appropriate structure loading cases. The load scenarios described in the following sections should be considered during design.

3.2.2 Structure Erection

Lifting a structure or components may generate greater stresses in the members of the structure than those induced by the in-service design loads (Figure 3-1). The load path may also be different and is determined by the chosen lift points. The designer should anticipate lifting limitations and provide attachment locations, as applicable, to control the load path.

(a)

(b)

Figure 3-1. Structure erection of (a) a tubular steel H-frame structure; and (b) a latticed steel guyed tower.

3.2.3 Loads Due to Wire Installation

Ground wires and conductors should be installed in accordance with the recommendations of IEEE Standard 524, *"IEEE Guide to the Installation of Overhead Transmission Line Conductors."* Ground wire and conductor stringing loads may be larger than the anticipated maximum intact design loads and are calculated from the stringing geometry and length of wire pull. Recommended minimum loads and load factors for installing ground wires and conductors include the following:

- For transverse and longitudinal components of wire tension, use a tension based on initial wire conditions at the lowest temperature that is expected to occur during stringing operations. Apply a minimum load factor of 1.5.
- For transverse wind loads, use a 3 psf (0.144 kPa) wind [35 mph (15.6 m/s)], the lowest temperature anticipated, and the maximum design wind span with a minimum load factor of 1.5.
- For dead-end conditions with pulling or tensioning equipment at ground level, use the vertical component of the pulling line, the maximum single vertical span, and a minimum load factor of 1.5.
- For intact wire conditions (ahead and back spans are attached to the structure), use the maximum design weight span and a minimum load factor of 2.0.

Sections 3.2.3.1 through 3.2.3.6 provide additional detail for specific wire installation loads that should be considered during the design phase.

3.2.3.1 Wire Tension Loads at Snub Structure At the ends of a wire pull, the wire passes over the stringing blocks and then downward to the pulling or tensioning equipment (Figure 3-2). A pulling line slope of at least three horizontal to one vertical is typically considered good practice. Because of this 3H:1V slope, the wire tension produces a vertical load at the location where the stringing blocks are mounted to the structure. A transverse load component may develop where the stringing blocks attach to the structure depending on the location of the pulling equipment. The transverse load component is a function of the angles made by the wire entering and leaving the stringing blocks due to the horizontal alignment of the tensioning equipment (Figure 3-3).

3.2.3.2 Wire Stringing Loads At structures in the middle of a wire pull, the wire passes over the stringing blocks. The transverse load component is a function of the angles made by the wire entering and leaving the stringing blocks due to the horizontal alignment of the transmission line. The vertical load component is a function of the conductor weight and the weight span (Figure 3-4).

Figure 3-2. Wire stringing operations and vertical and transverse load at snub structure.

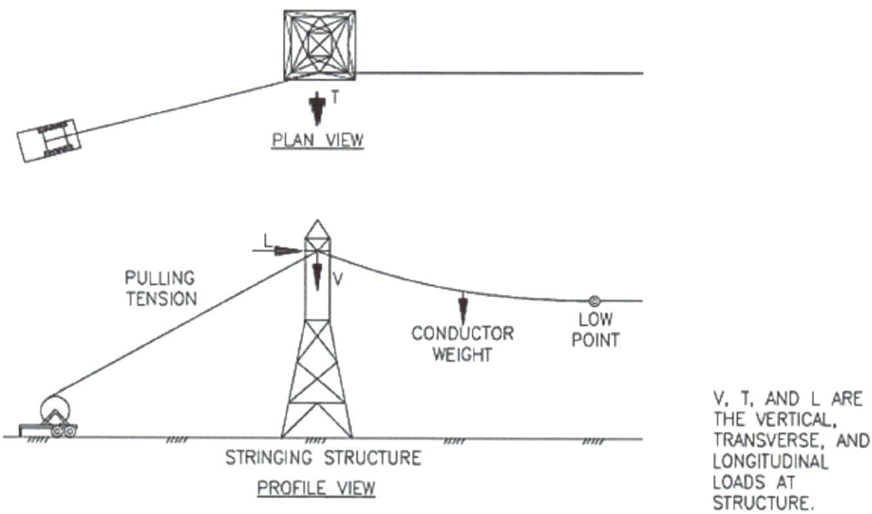

Figure 3-3. Vertical wire tension and wire weight components of load at snub structure.

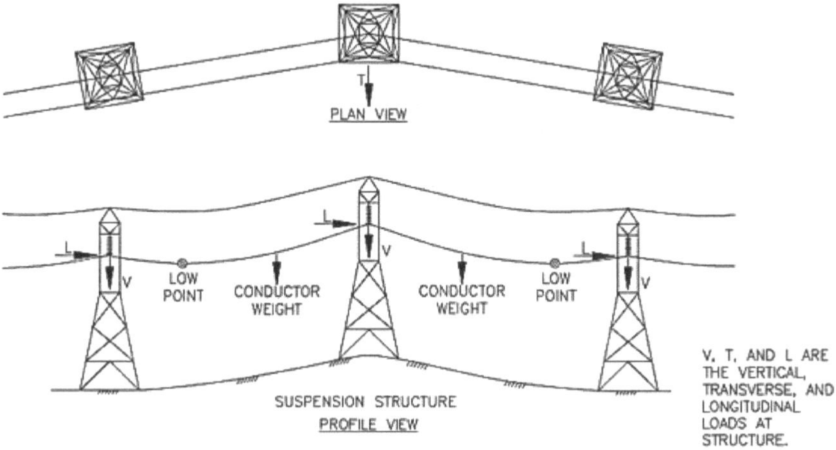

Figure 3-4. Intact stringing load conditions.

3.2.3.3 Bound Stringing Block Loads During tension stringing operations, the running board may jam in the stringing block. This can damage the structure when there is inadequate control or time to stop the pulling operation. Although a few utilities have designed suspension structures to resist such possible loads, a more practical solution is to control the stringing operation in accordance with IEEE Standard 524. See Figure 3-5 for a photo of wire stringing operations showing a running board near a stringing block.

Figure 3-5. Wire stringing operations.

3.2.3.4 Temporary Wire Attachment Loads As wire stringing progresses, the sequence of wire tensioning may require wire tensions to be transferred from tensioning equipment or temporary anchors to the structure. This could result in a configuration with some phases or subconductors installed and temporarily attached to a structure while other wires or phases are not in place or are being installed. Differential vertical loads and sagging tension loads should be incorporated in the design of the structure.

3.2.3.5 Vertical Load Increase in Hilly Terrain When pulling up a slope, the initial wire tension will increase by the unit weight of the conductor times the elevation change while the wire is in the blocks and before clipping-in and offsets are applied. This can severely increase the vertical load on uphill structures and must be considered in the design of the structure.

3.2.3.6 Temporary Guy Wire Loads Guy wires may be used to provide temporary support for suspension structures, crossarms, or other supports during stringing operations to control deflection and the load path. These temporary guy wires increase vertical loads on the structures. The design capacity of the guy wire system, structure, and structure components should exceed the anticipated temporary dead-end connection loads.

3.2.4 Maintenance Loads

Maintenance loads act on the structures as a result of scheduled or emergency inspection and/or replacement of all or part of a structure, ground wire, insulator, conductor, and conductor hardware system. An appropriate load factor should be applied when designing for maintenance loads. A load factor of 2.0 is generally recommended. Structure maintenance loads consist of the effects of workers on the structure and of load effects on adjacent structures due to temporary modifications, such as guying, to permit the repair or replacement of the structure being maintained. Refer to Section 3.3 for worker access and fall protection loads.

Common maintenance performed on a transmission line includes adjusting or replacing ground wires, conductors, insulators, and hardware. At times, it is necessary to remove the wires from their supports and lower them to the ground or transfer them to a temporary structure or temporary alternate location on the structure being maintained. The loads induced during these operations could exceed the original design loading of the structure or the adjacent structures.

Prior to lowering wires at one or more structures, the load effects should be considered. With level spans, the lowering of wires to or near the ground at one structure will increase the vertical loads at the adjacent structures and cause the insulators to swing toward the lowered line as the line tension attempts to equalize. This maintenance operation can create combined vertical and longitudinal loads, potentially overstressing the adjacent structures.

3.3 WORKER ACCESS AND FALL PROTECTION LOADS

During erection and maintenance, some structure members are loaded in flexure by the vertical weight of the workers. This access load should be applied only to members on which a worker is anticipated to climb or work. Normally this load is treated as an independent vertical load not less than 250 lb (1.1 kN), the approximate weight of a lineman with tools. A load factor of at least 1.5 is recommended.

Climbing devices used exclusively to support personnel, such as step bolts and ladders, as well as anchorages used for fall protection, may have additional strength requirements as stated in standards such as Occupational Safety and Health Administration (OSHA), ANSI C2 *National Electric Safety Code*, ANSI/ASSE Z359.1 *Fall Protection Code*, and IEEE 1307 *Standard for Fall Protection for Utility Work.*

Fall protection loads are created when workers, attached to an anchorage, fall from an elevated position. An anchorage is a secure point of attachment to which the fall protection system is connected (ANSI C2). The fall protection system must meet all OSHA 1910.269, ANSI C2, and other legislative requirements as applicable. In addition, IEEE 1307 provides guidance regarding loads and criteria for anchorages and step bolts. The design of the fall arrest system should be coordinated with operation and maintenance personnel.

3.4 WIND-INDUCED STRUCTURE VIBRATION

Wind-induced vibration can occur with or without the installation of conductors and shield wires and typically occurs in low-turbulence wind conditions. This is due to the fact that smoother wind flow leads to a more organized shedding of vortices from structural members. The frequency at which vortex shedding occurs for a component is proportional to wind velocity. When the wind velocity is such that the frequency of vortex shedding matches or is very close to the natural frequency of a component or system, a condition of harmonic resonance can occur. This mechanism tends to generate a large number of loading cycles and may result in fatigue issues or failure. The vortex shedding phenomenon is directly related to the shape of the member and can often be controlled by altering the profile of the member as seen by the wind. The amplitude of vibration at harmonic resonance is directly related to the amount of structural damping and can be mitigated by the addition of damping mechanisms. Appendix E contains additional information concerning wind-induced structure and arm vibration and oscillation of structures and members.

Transmission line structures can be subjected to dynamic forces caused by wind-induced vibration and conductor motion. These forces have the potential to initiate vibration in both the complete structure and individual members. In lieu of a comprehensive engineering analysis considering local weather conditions and individual structure member characteristics, the effects of structure vibration may be reduced by one or more of the following methods: hardware assembly design, structure design, member detailing, connection detailing, wire vibration dampers, and erection methodology.

When conductors or ground wires are not installed immediately after the tubular steel pole arms and crossarms are installed, owners should install mechanisms to minimize potential wind-induced oscillation. Examples of these mechanisms include the installation of internal or external damping devices, internal cables, weights, temporary tiebacks to a fixed point, and insulator assemblies with travelers (ASCE 2011). The structure designer should be consulted to determine what measures, if any, should be used for each specific circumstance.

3.5 WIRE GALLOPING LOAD CONSIDERATIONS

Galloping (high-amplitude, low-frequency wire motion) is a dynamic load that can occur given the right aerodynamic conditions of low-turbulence wind blowing across wires with or without a relatively light ice coating (Figure 3-6). Galloping events are unpredictable and can occur in one or several phases or spans. Factors influencing the onset of galloping include wire orientation, diameter, shape, weight, frequency, damping, and span length. However, galloping is more prevalent in flat terrain when wind speeds are between 10 and 20 mph (16.1 and 32.2 kmph), with wind blowing fairly normal to the wires, with an uneven ice coating. Because of this, random galloping occurrences may be more frequent than heavy ice loading events. Therefore, the effects of galloping should be considered in the design of the line when these conditions exist.

The vertical amplitude of the galloping wire can reach or exceed the wire sag, although most galloping amplitudes are less than 1 m (Den Hartog 1932, Davison et al. 1961, EPRI 1979, Havard and Pohlman 1980, Rawlins 1981).

The effects of galloping may cause electrical and structural problems such as

- Flashovers or clashing of wires that lead to temporary or permanent outages due to the reduction of spacing between phases or a phase and a ground wire (Farr 1980, REA 1980);

Figure 3-6. Conductor galloping.

- Permanent elongation of conductors and ground wires caused by dynamic wire tensions in the inelastic range. This leads to additional sag which may infringe on allowable ground clearance (Anjo et al. 1974, Richardson 1986);
- Excessive wear, fatiguing, and failure of ground wires, conductors, and associated hardware and insulators of suspension and dead-end assemblies (EPRI 1979); and
- The collapse of structural components and systems (Baenziger et al. 1993a, b; White 1979).

3.5.1 Wire Galloping Loads

Galloping wires can produce large vertical and longitudinal loads at supports. Theoretical studies indicate that tensions at dead-ends can vary by ±60% and the vertical loads at support points can vary by ±30%, with the magnitude depending on many factors (Brokenshire 1979, Gibbon 1984, Richardson 1986, CIGRÉ 2007). Measurements on actual galloping lines have found tension changes at dead-ends, cycling between 80% to 140% of the static tension (Anjo et al. 1974). It should be noted that cycling of vertical loads at support points, which were measured to be of the same magnitude as the tension changes, may not be visibly evident if the support point is rigid (Anjo et al. 1974). These vertical and tension loads may cause structure or structure element failure, such as ground wire masts (White 1979).

3.5.2 Galloping Mitigation

Several methods of galloping mitigation have been implemented. For new line construction, utility experience has shown that twisted pair conductors can reduce the occurrence of galloping. For existing lines with round conductors, galloping mitigation measures include detuning pendulums, interphase spacers, airflow spoilers, dampers, and rotational weights. These and other alternative measures and devices have been evaluated in field investigations (EPRI 1979, Havard and Pohlman 1980, Havard et al. 1982, Nigol and Havard 1978, Pohlman and Rawlins 1979, Whapam 1982). Experiences with mitigation devices indicate varying degrees of success.

Although increasing the vertical and horizontal spacing between wires may eliminate flashovers or clashing, this solution may require longer arms, taller and larger structures, and larger foundations (Boddy and Rice 2009). Additionally, increased wire spacing will not eliminate the other potential structural problems associated with galloping.

3.6 EARTHQUAKE LOADS

Transmission structures typically need not be designed for ground-induced vibrations caused by earthquake motion. Historically, transmission structures have performed well in seismic events as documented in industry publications, such as ASCE Technical Council on Lifeline Earthquake Engineering (TCLEE) and Earthquake Engineering Research Institute (EERI) reconnaissance reports. Decades of experience with lines of all configurations support the fact, that few, if any, failures from inertial loads are seen on transmission structures after an earthquake event. Computer modeling of both latticed towers and tubular steel structures has shown that the structure loadings caused by extreme wind, ice, and unbalanced wire tension loads exceed the loads caused by earthquake events (Riley et al. 2002). Transmission lines are a complex system of structures, overhead wires, and insulators. The varying structure types and heights, and cable span lengths and sags, interconnected with flexible insulators along the line, result in significantly different lower natural frequencies between the structures and wires. The advantage of this system is the low relative mass of the structures and the ability of the cable system to dissipate dynamic energy. Structure failures during a seismic event have been caused by geotechnical effects, such as landslides, liquefaction, and lateral spreading.

Thus, the inclusion of seismic inertial loads typically will not control the design of a transmission structure. The traditional extreme loads, as provided in this manual, are adequate to obtain the structural capacity to mitigate the effects of earthquake inertial loads. The following sections provide guidance when seismic effects are considered in the design of a transmission line.

3.6.1 Seismic Hazards

Transmission lines are long continuous structural systems that may encounter a variety of seismic hazards along potential or established routes. While the design of transmission structures will typically not be controlled by seismic loads, structural damage could occur as a result of secondary ground or terrain effects around a structure. When designing a transmission line in seismic regions, the following hazards should be considered by designers to minimize potential transmission structure damage:

- Induced seismic loads on structures subject to strong ground motion: The transmission structure is part of the line system, which consists of the structures and supporting wires interconnected by the insulators. The complexity of this system can vary with structure types, heights, leg configurations, span lengths, sags, wire configurations, insulator arrangements, and terrain. To understand the seismic inertial load on a transmission structure, these parameters need be included in a structural finite-element model of the transmission line system to obtain a representative inertial load. This type of detailed analysis is typically not performed because of the lack of structure failures caused by inertia loads during earthquake events. For new unique, nontraditional, structure configurations and/or materials, the need for earthquake inertial loads should be considered.
- Ground rupture and ground movement: Structures built across an earthquake fault zone could be damaged as a result of permanent ground displacement due to the fault. Adequate conductor sag and slack, insulator type, and configuration should be considered to accommodate large horizontal displacements when siting a transmission line that crosses a fault zone, particularly at a lateral strike/slip fault.
- Liquefaction: Seismic shaking can result in ground subsidence (settlement and lateral spreading) under certain soil and groundwater conditions. Some soils can liquefy during an earthquake and cause structures that are founded on such materials to experience

differential foundation settlement and/or large horizontal movement resulting in structural damage or failure.

- Landslides: Seismic shaking can induce landslides which could either undermine a structure's foundation or cause slide debris such as rock falls to impact a structure. The terrain and geology along the transmission line route should be studied to determine where these hazards may occur.

3.6.2 Siting and Geotechnical Assessment

Seismic design relative to a transmission line design is normally limited to a geotechnical earthquake hazard assessment along the line route. Geotechnical work for a specific project should include an evaluation of areas susceptible to liquefaction and landslides as well as likely fault rupture zones. Where seismic activity along a transmission line route is a potential, a qualified geologist or geotechnical engineer familiar with the seismicity of the area should be part of the project routing review to provide input for routing decisions.

3.7 SUMMARY OF ADDITIONAL LOAD CONSIDERATIONS

Table 3-1 summarizes the recommended load factors that are referenced within this chapter and associated appendixes.

Table 3-1. Summary of Additional Load Considerations

Recommended load case	Section	Weather conditions	Recommended load factors: Longitudinal	Transverse	Vertical
Line security loads	3.1.4	See Appendix I			
RSL method					
EPRI method					
BPA method					
BWL method					
Structure erection loads	3.2.2	—	1.5	1.5	1.5
Wire installation loads					
Wire tension loads at snub structure	3.2.3.1	Lowest anticipated ambient temperature at time of installation; wind = 3 psf	1.5	1.5	1.5
Intact wire loads at tangent structure	3.2.3.2		1.5	1.5	2.0
Bound block loads	3.2.3.2		2.0	2.0	2.0
Maintenance loads	3.2.4	Lowest anticipated ambient temperature at time of installation; wind = 3 psf	2.0	2.0	2.0
Worker access and fall protection loads	3.3	—	1.5	1.5	1.5
Wind-induced structure vibration	3.4	See Appendix E			
Wire galloping	3.5	—	—	—	—
Earthquake loads	3.6	—	—	—	—

CHAPTER 4
WIRE SYSTEM

4.0 INTRODUCTION

A wire system includes conductors, overhead shield wires, and other cables connected to the structure. Structural wires, such as guy wires, are not considered part of the wire system. It is necessary to understand the tension in the wire system as well as the longitudinal, vertical, and transverse loads these wires impose on the structure attachment points. Wire tensions vary with

- Electrical loading: current and resistance;
- Weather loading: ice, wind, and temperature;
- Wire condition: tensions at initial, final after creep, and final after load;
- Structure support flexibility: structure type and insulator configuration;
- Construction and maintenance: stringing and erection techniques; and
- Span characteristics: span lengths and elevation changes.

The effects listed above alter the temperature, length or support point, and therefore tension, of each wire. Wire tensions temporarily decrease as wire temperature increases and thermally expands (elastically lengthens) the wire. Conversely, tensions increase with decreased temperatures. Similarly, tensions decrease as wires creep (plastically lengthen) with load and time. External loading of ice and wind will also increase wire tension. These and other loading issues will cause any given wire to experience a range of tensions. The wire tensions must be contained within limits to ensure the viability and performance of the wires and of other components of the wire systems.

The wire tensions directly affect

- Longitudinal loads applied to the strain and dead-end structures,
- Transverse loads at all line angles,
- Factored longitudinal design loading or Residual Static Load (RSL) to resist cascades (See Chapter 3),
- Vertical loads at structures with a vertical angle (VA) (See Figure 4-1), and
- Foundation, guy wires, and anchors.

Loads per unit length of conductor or ground wire have been discussed in Chapter 2. This chapter discusses the manner in which the wire system is affected and the assumptions that may be used for determining the loads at the structure attachment points.

4.1 TENSION SECTION

The wire system is normally broken down into tension sections. A tension section is a portion of conductor or ground wire strung between dead-end points, such as between Points A and E in Figure 4-1. Wire tension only affects structures where a horizontal or vertical angle occurs in the tension section and where wires terminate. If the wires at the supporting structures do not have a horizontal angle (Points B and D in Figure 4-1), then the transverse loads are not affected by wire tensions. For structures with a vertical angle (Location B in Figure 4-1), the vertical and

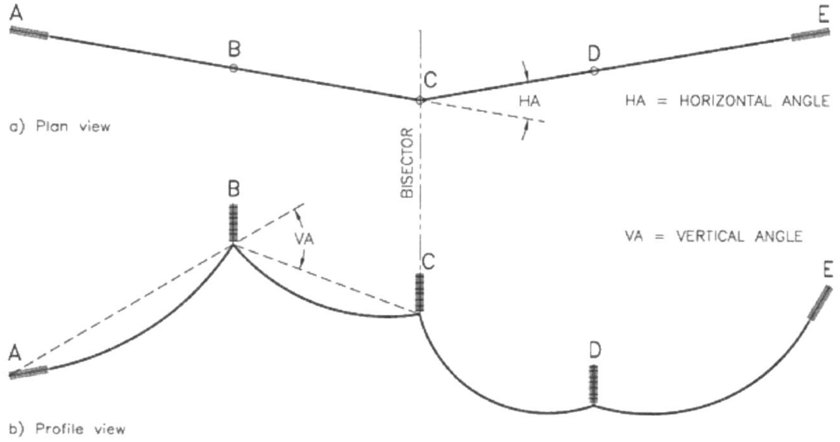

Figure 4-1. (a) Plan; and (b) profile of tension section.

longitudinal load could be affected by the wire tensions depending on the geometry of the spans.

Intermediate wire attachment points B, C, and D between the dead-end points are presumed to have some longitudinal flexibility. This flexibility, or freedom to move, equalizes the horizontal components of tension for various loading events. Longitudinal flexibility in the wire system is provided by two possible sources: insulator configuration and structure type. At Points B, C, and D, suspension insulators allow more wire movement than post insulators or overhead ground wire clamps. Poles or H-frames exhibit more longitudinal flexibility than more rigid lattice towers or guyed structures. The importance and complexity of the wire system longitudinal flexibility is discussed in Section 4.4.1.

4.2 WIRE CONDITION

Wires, especially conductors, are subject to permanent elongations over their lifetime in service. They are said to be in their "initial" condition if they are new wires and within a few hours of installation. The wires are in their "final after creep" condition if they have been in service for several years and have experienced permanent elongation over time under normal operation. The creep process slows considerably with time. Estimates of creep are usually based on a 10-year period. Therefore, the wire will spend most of its life at a condition close to what is commonly called "final after creep." The majority of creep elongation occurs in the initial one or two years following the conductor installation.

Assumptions regarding wire elongation and "initial" or "final" conditions are based on wires of aluminum, steel, and/or copper, commonly used for electrical transmission lines. In recent years, high temperature–low sag conductors have been developed, which have different material characteristics.

The wires are in their "final after load" condition if they have been permanently elongated by a high tension due to the maximum design load, such as an extreme load from ice, wind, or a combination of both as described in Chapter 2. Permanent elongation from heavy load can be determined more precisely than elongation from creep if future load function or load history can be accurately predicted. However, since the selection of a weather-related load (used in the calculation of wire tension) is a design assumption (future estimation of applied load), the permanent elongation calculated is also an estimation. The magnitude of the "final after load" cases, selected for design, is intended to be an upper limit and should be evaluated using computational methods available in typical design software. Since "final after load" depends on extreme weather cases such as extreme wind or heavy ice, it is possible that this load condition may never occur over the life of the line.

The total permanent elongation of the conductor under the impact of creep or load will depend on the time loading history of the conductors. For example, if a severe weather load occurs very early in the life of the line, the permanent load elongation will cause the everyday conductor tension to decrease.

An accepted practice in the industry is to disassociate creep and load permanent elongation and assume them to be independent and not additive. Compared to their initial values, tensions are lower for "final after creep" and "final after load" conditions because the wires have permanently elongated. A situation where a "final after creep" tension is lower than the corresponding "final after load" tension is sometimes referred to as "creep is a factor" or "creep controls." For certain design aspects such as aeolian vibration, it is best to consider both wire states. An alternate method has been offered by CIGRÉ Task Force B2.11.04 (2004) for tensioning of wires with respect to aeolian vibration.

In addition to sag-and-tension related wire conditions such as initial and final, conductors, ground wires, and other components of the wire subsystem can deteriorate. Some electrical, environmental, and other physical impacts on wire condition are

- Lightning strikes,
- Wire motion fatigue,
- Corrosion of steel strands,
- High electrical loading,
- Impact damage,
- Electrical flashovers, and
- Hardware fatigue.

4.3 WIRE TENSION LIMITS

It is critically important to protect the wire system because failure of a wire, or any components in series with the wire, will impose longitudinal loads on the structure system that may exceed design loads causing failure. Wires are normally sagged to perform within certain design limits. Limits on everyday tensions or on catenary constants under initial and/or final conditions are normally given to avoid or to minimize the potential for wind vibration damage.

For tension limits, legislated requirements such as NESC or IEC 60826 along with wire manufacturer recommendations should be followed. Many utilities specify more stringent tension limits. The everyday wire condition should be specified as "final after creep" and not "final after load" since the "final after creep" generally has a higher tension magnitude for the reasons discussed in Section 4.2. For mitigation of wire vibration, the tension associated with "final after creep" is typically considered a prudent design practice as the wire will be in that tension condition for a majority of its service life.

Designing using "final after load" may result in a wire enduring higher tension for its service period until the heavy load occurs. This can cause unforeseen vibration problems due to a higher tension. Wire vibration mitigation devices should be installed with appropriate consideration of wire tensions, span lengths, and local weather conditions.

When extreme loads do occur on the wire systems, all the components in series within these systems are highly stressed and must have adequate remaining strength. Therefore, it is recommended that the maximum tensions caused by the factored extreme loads never exceed 70% to 80% of the rated tensile strength of the wire.

4.4 CALCULATED WIRE TENSION

4.4.1 The Ruling Span Method

Assuming that the horizontal components of tension, T_H, in all the spans of a tension section are the same when the spans are in relatively flat terrain, then the entire tension section can be represented as a single equivalent "ruling" span (Thayer 1924). The wire horizontal tension, T_H, for a given set of temperatures and external loads may be determined by subjecting the ruling span to those conditions.

The ruling span method implies that the same unit load is applied on all the spans of the tension section. To make this assumption valid, the intermediate support points must have sufficient longitudinal flexibility to allow this equalization of tensions. This method is valid for wire sag and tension calculations where span lengths are similar. Rugged mountain terrain and sizeable span length differences reduce the accuracy of the ruling span model. The ruling span is an approximation that has limitations (IEEE 1999). The ruling span is computed as follows:

$$\text{Ruling Span} = \sqrt{\frac{S_1^3 + S_2^3 + S_3^3 + \ldots + S_n^3}{S_1 + S_2 + S_3 + \ldots + S_n}} \tag{4-1}$$

where $S_1, S_2, S_3, \ldots, S_n$ is the individual span lengths (horizontal projections) between dead-end or strain structures.

When correctly applied, the ruling span method [Equation (4-1)] enables the stringing and sagging-in of a line section (i.e., between dead-ends) of unequal spans in flat or hilly terrain so that the horizontal tensions in each span will be equal as designed.

The ruling span method relies on the ability of the wire connection points to move longitudinally without restriction as a means to equalize tensions between spans. Calculated tensions may not be as accurate for conductor spans supported by rigid post insulators or for ground wires supported by rigid clamps. They may also not be as accurate if suspension insulators are not sufficiently free to move in the longitudinal direction (e.g., at support points with substantial horizontal and vertical line angles). In such cases, the structural analysis options described in the next subsections are more appropriate.

The ruling span method is not applicable to the calculation of unbalanced longitudinal loads caused by restricted suspension points, uneven ice on adjacent spans, or other span-specific disturbances.

4.4.2 Structural Analysis of a Single Tension Section

If there is no significant interaction between consecutive tension sections caused by the longitudinal displacement of their supporting structures, then the tensions in the various spans of a tension section can be determined by modeling that tension section as a wire system with appropriate support conditions. Suspension supports can be modeled as cable elements or swinging rods. Post insulators can be modeled as small, cantilevered beams or longitudinal springs with appropriate longitudinal flexibilities. The model is then analyzed by accepted structural analysis methods that account for the longitudinal displacements of the wire attachment points while incorporating the corresponding changes in wire tensions.

This type of analysis is capable of handling unbalanced ice and/or wind loads and will produce more accurate tension results than the ruling span method.

4.4.3 Structural Analysis of Multiple Tension Sections

If there is significant interaction between consecutive tension sections caused by the longitudinal displacement of their supporting structures, then those tension sections can be analyzed as a single structural system, which includes all the wires in all the spans as well as detailed structural models of all supports. This rigorous approach produces a more accurate analysis but, because of its complexity, is normally only justified in special situations.

4.4.4 Computational Methods

Modeling of a tension section or multiple tension sections depends on the wire system and support modeling methods and assumptions since they operate in concert under various load cases. For 2D models where the ruling span, actual wind/weight span, attachment displacement, and individual conductor line angle may all, or in part, be held constant under different load cases for computational simplification, the results may be divergent and varied. This is because these effects may be too critical to ignore, especially if the tension sections have significant combinations of line angles, permanent structure deflections, and/or elevation differentiation. In 3D models that incorporate lateral and longitudinal movements for conductor, support hardware, and structure, the computational results reflect a more accurate result than 2D models.

4.5 LOADS AT WIRE ATTACHMENT POINTS

4.5.1 Wire Unit Loads

A length of wire, as received from the manufacturer, possesses certain physical and electrical properties making it suitable for use on a transmission line. Electrical properties, such as the area of aluminum, or other conductive material, allow it to transfer current between one station and another. The physical and dimensional properties of that aluminum, in connection with steel or other strength components in the conductor, give resultant diameters, weight, and tensile strength. These basic wire properties of diameter, weight, and tensile strength are associated with the major transverse, vertical, and longitudinal loads imposed on their supporting structures.

The basic loads applied to wires can be divided into vertical (weight), transverse (wind), and longitudinal (tension) components (Figure 4-2). Vertical load (applied downward) and horizontal or transverse loads (applied perpendicular to the line) are calculated on a unit-of-length basis. These load components can be resolved into a resultant vector load. Longitudinal load (applied along the line), due to wire tension, is not typically considered as a unit load.

Bare wire will have the dimensional properties and stress versus strain characteristics as given by the supplier, yet wire in service is exposed to varying weather and construction and maintenance loads, which cause the unit loads to change with environmental conditions and loading history. Horizontal load is a function of wind pressure as discussed in Chapter 2 applied to the projected area of the wire, which may increase with accumulated ice or snow. Vertical load is a function of the bare wire weight coupled with that of any ice or snow. The Vertical Unit Load, w_v, and Horizontal Unit Load, w_h, are calculated using Figure 4-2. These unit wire loads vary with loading history and weather conditions as described. All significant load cases should be considered in structure design.

4.5.2 Using Wind and Weight Spans

Vertical and transverse unit wire loads, w_v and w_k, are multiplied by the applicable wind or weight span to obtain the wire loads applied to the

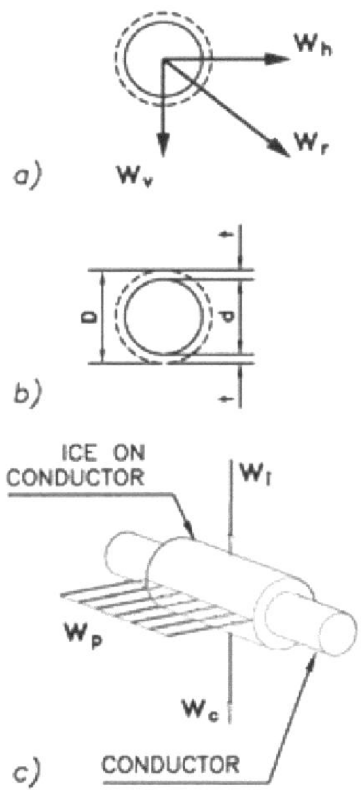

$w_r = \sqrt{w_v^2 + w_h^2}$

$w_v = w_c + w_I$

$w_I = \frac{\pi}{4}(D^2 - d^2)\rho$

$w_h = d \cdot w_p$ or $w_h = D \cdot w_p$

$D = d + 2 \cdot t$

w_r : UNIT RESULTANT WIRE LOAD
w_v : UNIT VERTICAL WIRE LOAD
w_h : UNIT TRANSVERSE WIRE LOAD
w_c : UNIT WEIGHT OF WIRE
w_I : UNIT WEIGHT OF ICE COATING (WHERE APPLICABLE)
w_p : WIND PRESSURE
d : WIRE DIAMETER
t : ICE THICKNESS
ρ : ICE DENSITY

Figure 4-2. Unit wire loads.

structures. At tangent locations, such as Points B and D in Figure 4-1, the transverse and vertical structure loads, L_T and L_V, can be determined as

$$L_T = \text{load factor} \times w_h \times \text{wind span} \tag{4-2}$$

$$L_V = \text{load factor} \times w_v \times \text{weight span} \tag{4-3}$$

where wind span is the length of wire between midspan points in the adjacent spans. In traditional (2D) calculations, the wind span was calculated as one half of the horizontal projections of the adjacent spans as shown in Figure 4-3. Weight span is the length of wire between the low points in the adjacent spans. In traditional (2D) calculations, the weight span was calculated as the horizontal distance between the low points in the adjacent spans as shown in Figures 4-3 and 4-4.

Three formulas are presented to locate the low point of a span (Winkelman 1959): an approximate equation [Equation (4-4)] or more precise catenary equations [Equations (4-5) and (4-6)]. It must be noted that Equation (4-4) should not be used if the difference in support elevations (B) is greater than approximately 20% of the span length (S).

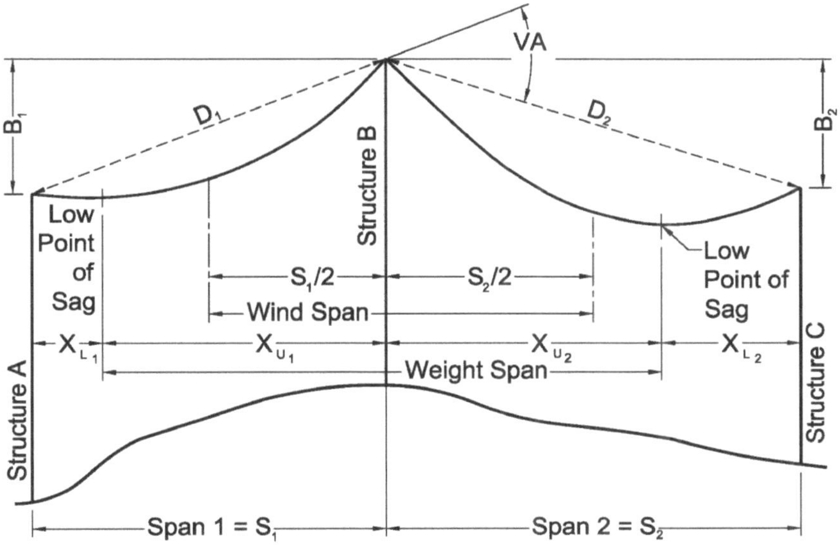

Figure 4-3. Wind and weight spans.

Approximate Formula. The position of the low point of sag, X_L, is approximated by the following formula:

$$X_L = \frac{S}{2 \times \text{midspan sag}} \times (\text{midspan sag} - B_i/4)$$

$$\text{sag}_m = \text{midspan sag} = (S_i \times D_i)/(8 \times C) \tag{4-4}$$

$$D_i = \sqrt{S_i^2 + B_i^2}$$

where subscript *i* denotes the span being considered from Figure 4-3.

Calculating the weight span for a particular wire loading requires determining the equilibrium configuration of the wire for that loading.

These low points can also be determined using the hyperbolic equations for a catenary shown in Equations (4-5) and (4-6). As terrain difference increases (slope angle VA becomes large), the calculated arc length of the wire *l* [Equation (4-7)] between the low point and the support should be used for calculating L_V using Equation (4-10).

$$X_L = \frac{S}{2} - C \times \sinh^{-1}\left(\frac{B/2}{C \times \sinh\left(\frac{S}{2 \times C}\right)}\right) \tag{4-5}$$

$$X_U = \frac{S}{2} + C \times \sinh^{-1}\left(\frac{B/2}{C \times \sinh\left(\frac{S}{2 \times C}\right)}\right) = S - X_L \tag{4-6}$$

$$l = C\left(\sinh \frac{X}{C}\right) \tag{4-7}$$

$$C = T_H / w_r \tag{4-8}$$

where

- C = Catenary constant, which is the ratio of horizontal tension to the unit wire load,
- S = Span length,
- B = Difference in elevation of supports,
- D = Straight-line distance between the supports,
- X_L = Distance from low point of sag to lower support,
- X_U = Distance from low point of sag to upper support,
- l = Length of wire from low point to support [$X = X_L$ for lower support and $X = X_U$ for upper support in Equation (4-7)],
- T_H = Horizontal component of tension, and
- w_r = Unit wire load for the desired load case.

Equations (4-4) to (4-8) assume that there is no blow-out of the spans, which is only true if w_r equals w_v ($w_k = 0$).

In such a case, the cables in Figure 4-3 are in a vertical plane. With some wind, the planes of the cables rotate around the straight line between the supports in the direction of w_r. However, the general aspect of Figure 4-3 remains valid, but with different sags and locations of the low points. These sags and low points can still be located using Equations (4-4) to (4-8) if one replaces the catenary constant of Equation (4-8) by $T_H/_{wv}$.

At line angle locations, such as Structure C in Figure 4-1, the horizontal component of wire tension in the adjacent spans results in transverse load.

This additional transverse load is calculated in Equation (4-8) and should be added to the transverse load in Equation (4-2)

$$L_T\text{-Angle} = 2 \times T_H(\sin 0.5\text{HA}) \quad (4\text{-}9)$$

where T_H = horizontal component of tension, and HA = horizontal line angle.

A good approximation of the vertical load, L_V, can also be obtained by Equation (4-10) for inclined spans versus wire arc length calculation.

$$L_V = \text{factored unit vertical wire load} \times \text{wind span} + T_H \times 2 \times \tan[0.5 \times \text{VA}] \quad (4\text{-}10)$$

where VA = vertical line angle.

4.5.3 Weight Spans on Inclined Spans

The low point of sag for structures with equal wire attachment elevations occurs at the midpoint of the span. The most simplistic calculation of weight span occurs when three consecutive structures support the wire at the same elevation. In this case, the weight span applied to the center structure equals half the sum of the two adjacent span lengths.

Where wire attachment elevation points are not equal, the low point of sag will not coincide with the span midpoint. When slope increases, whether in terrain or in structure height, the low point will move away from the higher attachment point. The situation is compounded with changes in tension due to temperature- or weather-related events. With decreasing temperature, the wire length decreases, the wire tension increases, and the catenary curve created by the suspended wire becomes more flat. This also moves the sag low point (Figure 4-4).

In the "Maximum Sag" portion of Figure 4-4, it is clear to see the low points occurring in each span and between each structure. The weight spans for Structures B, C, and D are shown as the distance between the respective low points of sag. The "Minimum Sag" portion, however, shows that increased wire tension flattens the curve of the wire catenary and can result in there being no low point (or belly) of sag within the actual span. Projected low points in the figure are indicated by dotted lines.

Wire tension changes may increase or decrease the weight span of a given structure depending on terrain. In the case of Structure D, the projected low points for minimum sag are beyond the structure from both adjacent spans. Structure D therefore has a negative weight span, meaning vertical loads through wires are upward rather than downward and are trying to lift the structure.

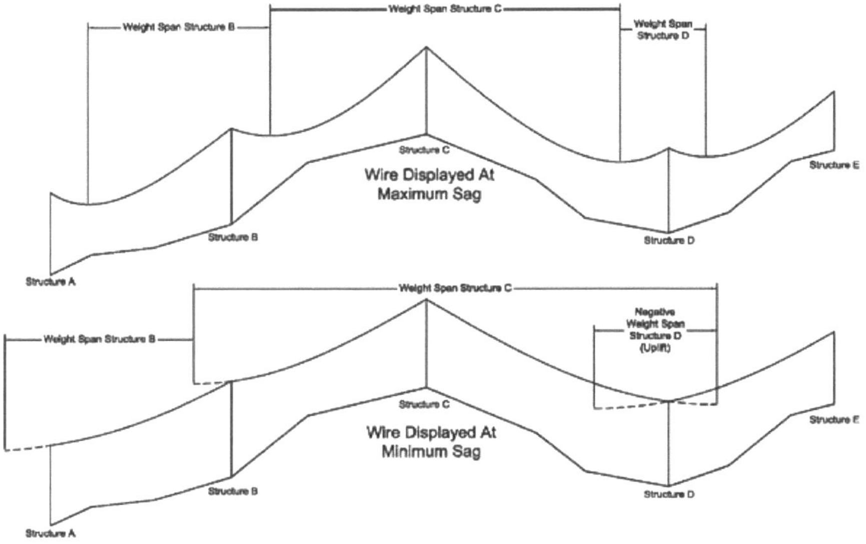

Figure 4-4. Weight span for inclined spans based on varying sags.

Transmission designers must account for this varying weight span and its effect on structure loading. Structures subjected to a negative weight span (or uplift) under certain loading conditions should use insulator and hardware assemblies designed to resist these vertical forces. These upward vertical loads must also be considered in the structure and foundation designs.

Conversely, structures must be designed to support increased vertical loads as the weight span is lengthened when wire tension decreases. Knowing the range of possible weight spans for a structure allows the application of the proper range of vertical loads in the structure and foundation designs.

Equations (4-2) and (4-3) can be used to calculate weight load distributions based on tension and elevation conditions.

4.5.4 Weight Span Change with Blow-Out on Inclined Spans

Extreme transverse winds on inclined spans can result in large horizontal conductor displacement, commonly known as blow-out. This can produce changes to the weight spans at the supports. The location of the low point of the sag (or projected low point) can shift dramatically in the span. When slopes exceed approximately 20%, the weight span at the upper support point can approach double that calculated by methods which ignore the blow-out, and may even exceed the factored construction and maintenance loads. Furthermore, the reduction of weight at the lower point may lead to excessive swing of the insulator strings and result in inadequate electrical clearances to the structure. Also, the net vertical force may be an uplift that can collapse the crossarm. Some line design computer programs include calculation of the shift of weight spans with blow-out of inclined spans. The methods discussed in Section 4.5.1 provide a means to determine if this condition exists [Equations (4-5) and (4-6)].

If the span inclination exceeds about 25% and extreme winds are to be expected, more precise calculations may be considered or else conservative vertical load values may be applied to the structures at the upper and lower support points. These calculations can be made analytically (Keselman and Motlis 1996, 1998) or with a finite-element computer program (Peyrot 1985) by breaking the spans into short cable elements, with each element responding individually to a different local wind incidence. It is important to note that the analysis of this severe blow-out problem is very dependent on the horizontal angle at which the wind strikes the span and on the vertical angle of approach of the wind. Both of these angles are likely to vary considerably and randomly in rough or mountainous terrain where steeply inclined spans are to be found. Deviations from the orthogonal can greatly increase these distortions of the wire systems and increase or decrease the expected weight span changes. Thus, if a serious blow-out problem is anticipated, there is even greater justification for a conservative approach to the strengths of the upper and lower structures.

Weight spans of structures vary with temperature due to the wire length changing with temperature. The weight spans of a structure with higher wire attachment elevations than the adjacent structure will increase as the wire temperature decreases. This is due to the low point of the sag moving away from the higher attachment points toward the lower attachment points (downhill) as the wire contracts due to the temperature change. The opposite is true at structures that are lower than adjacent structures. Therefore, in no-ice areas the largest vertical load, which can occur at a higher structure, is generally caused by the coldest temperature or sometimes by wind blow-out. Similarly, the coldest temperature or the wind blow-out can cause uplift and insulator swing problems at lower structures.

In icing areas, the weight under ice should be used to calculate the vertical load with Equation (4-3). For higher towers, it will almost certainly be found that the iced weight span is substantially less than the cold bare-wire weight span.

Equations (4-2) and (4-3) can be used to determine design loads on a new family of structures intended to have transverse and vertical capabilities based on assumed maximum (allowable) wind and weight spans.

When these structures are spotted, their ability to carry their design loads at a particular location is simply checked by verifying that the actual (as spotted) wind and weight spans are less than the allowable values. In icing areas, the fact that iced weight spans are generally shorter than cold bare-weight spans can be used to advantage by specifying shorter allowable weight spans under ice than under bare cold.

The concept of allowable wind and weight spans is extremely useful when spotting new lines, especially with families of standardized structures. However, for the design of custom structures at specific locations, for the checking of existing lines, or for parametric studies for possible upgrading or reconductoring, there is no need to be concerned with approximations in the wind and weight spans approach if the loads are computed by a structural analysis method that accounts for the 3D behavior of the wire system.

4.5.5 Centerline Horizontal Angle versus Wire Horizontal Angle

When the wire configuration changes either by rolling phases or arm length changes, the wire angle of one or more phases can vary from the centerline angle. For design purposes, the centerline angle is often employed as the wire horizontal angle for loading purposes. For a line design that utilizes the same structure geometry, this is a valid general assumption. However, it is important to remember that the loading should be calculated using the wire angle rather than the centerline angle when the two differ from one another. A given structure may have several different wire angles depending on the structure geometry and wire arrangement. For these structure locations, the assumption of the centerline angle representing the wire loads may not be sufficiently accurate for the intended purposes.

Examples of this include

- Vertical to horizontal phase rolling, or phase transpositions;
- Vertical to delta configuration;
- Other structure type changes with different phase geometries; and
- Alignment convergence.

CHAPTER 5
EXAMPLES

5.0 LATTICED SUSPENSION TOWER LOADS

This example shows calculations for wire and structure loads. The loads are based on the tower shown in Figure 5-1, Tables 5-1 through 5-3, and the design and wire data listed below. Notation within this chapter has been kept consistent with other chapters where possible. For convenience, the coordinate system and variable definitions below only apply to this chapter. Calculation results have been rounded to simplify presentation in the tables. Nothing is to be inferred from the rounding methods used in this chapter.

Structure Coordinate System:

- Vertical axis: Local axis that is parallel with the direction of gravitational force.
- Longitudinal axis: Local axis that is parallel with the general direction of the transmission line conductors and perpendicular to the vertical axis.
- Transverse axis: Local axis that is perpendicular to the plane formed by the vertical and longitudinal axes.

Loading nomenclature:

- V = Structure load that is parallel with the vertical axis.
- L = Structure load that is parallel with the longitudinal axis.
- T = Structure load that is parallel with the transverse axis.
- H = Horizontal component of wire tension.

5.0.1 Design Data

Location: Utah
MRI = 100 years
90 mph extreme wind speed
0.25 inch ice, 40 mph concurrent wind
Ruling span = 1250 ft
Wind span =1500 ft
Weight span = 1800 ft
Line angle = 5°
Length of insulator assembly = 6 ft
Weight of insulator assembly = 200 lb
Weight of shield wire assembly = 50 lb
No topographic effects, K_{zt} = 1.0
Wind pressures on wires for the two loading cases Wind and Wind at 30° *Yaw Angle* are calculated based on the wind span. For calculations of sags and tensions, the use of the ruling span is appropriate.

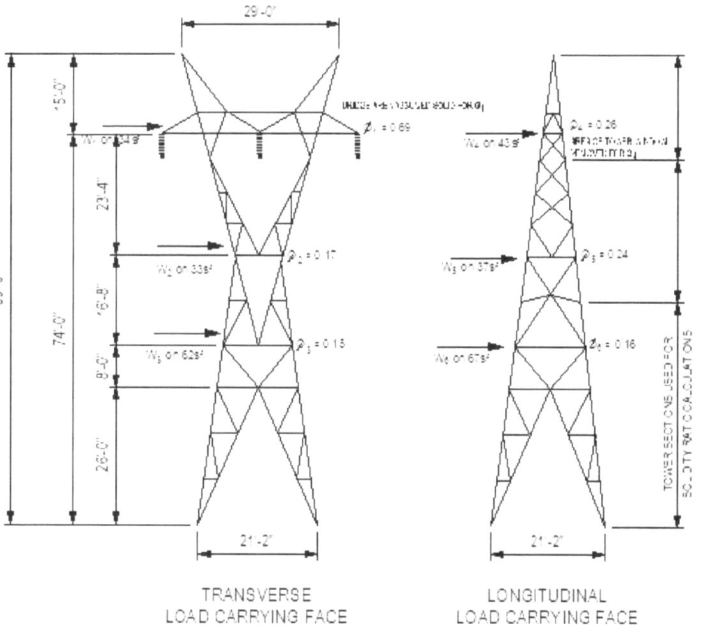

Figure 5-1. Suspension tower.

Table 5-1. Wire Data

Loading case	Temperature (°F)	Ice (in.)	Wind (psf)	954 kcmil 45/7 ACSR rail conductor (d = 1.165 in., w = 1.075 lb/ft)		7#8 aluminum clad steel shield wire (d = 0.385 in., w = 0.262 lb/ft)	
				Initial sag (ft)	Initial tension (lb)	Initial sag (ft)	Initial tension (lb)
Wind	60	0	16.1	43.5	8,530	38.8	2,917
Wind at 30°	60	0	12.1	42.1	7,396	37.0	2,472
Wind and ice	15	0.273	3.17	39.3	8,092	36.3	2,929
C&M	15	0	3	36.5	5,971	31.0	1,757
FC	30	0	0	37.4	5,622	31.4	1,628
No wind	60	0	0	39.6	5,305	33.0	1,550

Note: C&M = construction and maintenance, FC = failure containment load.

Table 5-2. Load Summary (kips)

Loading case	Shield wire			Conductor			Transverse wind on structure			Longitudinal wind on structure		
	V	T	L	V	T	L	F_{T1}	F_{T2}	F_{T3}	F_{L1}	F_{L2}	F_{L3}
Wind	0.5	1.0	—	2.1	3.1	—	1.2	2.0	3.9	—	—	—
Wind at 30°	0.5	0.8	—	2.1	2.4	—	1.5	2.0	4.0	0.9	1.2	2.3
Wind and ice	0.9	0.6	—	3.0	1.4	—	0.2	0.4	0.8	—	—	—
C&M	1.3	0.4	—	4.7	1.4	—	0.3	0.5	0.9	—	—	—
FC (broken wire)	0.3	0.1	1.6	1.2	0.2	3.9	—	—	—	—	—	—
FC (intact wire)	0.5	0.1	—	2.1	0.5	—	—	—	—	—	—	—

Note: C&M, construction and maintenance, FC = failure containment load, V = vertical load, T = transverse load, and L = longitudinal load.

Table 5-3. Weight Span Summary

		Upper tower		Lower tower	
Wire	C	Weight span (ft)	Difference (%)	Weight span (ft)	Difference (%)
Shield wire	Traditional	1652	+23%	424	−136%
	Section 5.1	2140	—	180	—
Conductor	Traditional	1608	+15	446	−45%
	Section 5.1	1884	—	308	—

Note: C-catenary constant or parameter of the catenary curve.

5.0.2 Extreme Wind (Chapter 2, Section 2.1)

For purposes of this example, the wind is assumed to be normal to the ahead span, back span, and to the structure. Note that the effective height of the wire system in this example was conservatively taken as the arm attachment height.

The structure is located on the perpendicular bisector of the line angle.

From the wind map (Figure 2-1 in Chapter 2), V_{100} = 90 mph, Exposure Category C.

5.0.2.1 Wind on Wires

Average wire height $z_h = \dfrac{3(74)+2(89)}{5} = 80$ ft from Section 2.1.4.3

$$K_z = 2.01\left(\frac{z_h}{z_g}\right)^{\frac{2}{\alpha}} = 2.01\left(\frac{80}{900}\right)^{\frac{2}{9.5}} = 1.21 \quad (2\text{-}3)$$

$$I_z = c_{\exp}\left(\frac{33}{z_h}\right)^{\frac{1}{6}} = 0.2\left(\frac{33}{80}\right)^{\frac{1}{6}} = 0.17 \quad (2\text{-}6)$$

$$B_w = \sqrt{\frac{1}{1+\frac{0.8S}{L_s}}} = \sqrt{\frac{1}{1+\frac{0.8(1500)}{220}}} = 0.394 \quad (2\text{-}8)$$

$$G_w = \left(\frac{1+4.6I_zB_w}{1+6.1I_z}\right) = \left(\frac{1+4.6(0.17)(0.394)}{1+6.1(0.17)}\right) = 0.64 \tag{2-5}$$

Wind pressure $$\frac{F}{A} = QK_zK_{zt}\left(V_{100}\right)^2 GC_f \tag{2-1a}$$

$$= (0.00256)(1.21)(1.0)(90)^2(0.64)(1.0)$$

$$= 16.1 \text{ psf}$$

5.0.2.2 Shield Wire Loads

V = vertical = w × weight span + hardware weight

= 0.262(1800) + 50 = 522 lb = 0.5 kips

$$T = \text{transverse} = \text{wind pressure} \times A + 2(\text{wire tension})\sin\frac{\Delta}{2}$$

$$= 16.1\left(\frac{0.385}{12}\right)(1500) + (2)(2917)\sin\left(\frac{5}{2}\right) = 1029 \text{ lb} = 1.0 \text{ kips}$$

5.0.2.3 Conductor Loads

V = vertical = w × weight span + hardware weight

= 1.075(1800) + 200 = 2135 lb = 2.1 kips

$$T = \text{transverse} = \text{wind pressure} \times A + 2(\text{wire tension})\sin\frac{\Delta}{2}$$

$$= 16.1\left(\frac{1.165}{12}\right)(1500) + (2)(8530)\sin\left(\frac{5}{2}\right) = 3089 \text{ lb} = 3.1 \text{ kips}$$

5.0.2.4 Wind on Structure

Two-thirds of the structure height $z_h = \dfrac{2(89)}{3} = 59.3$ ft from

Section 2.1.4.3

$$K_z = 2.01\left(\frac{z_h}{z_g}\right)^{\frac{2}{\alpha}} = 2.01\left(\frac{59.3}{900}\right)^{\frac{2}{9.5}} = 1.13 \tag{2-3}$$

$$I_z = c_{\exp}\left(\frac{33}{z_h}\right)^{\frac{1}{6}} = 0.2\left(\frac{33}{59.3}\right)^{\frac{1}{6}} = 0.18 \quad (2\text{-}6)$$

$$B_t = \sqrt{\frac{1}{1+\frac{0.56z_h}{L_s}}} = \sqrt{\frac{1}{1+\frac{0.56(59.3)}{220}}} = 0.932 \quad (2\text{-}7)$$

$$G_t = \left(\frac{1+4.6I_zB_t}{1+6.1I_z}\right) = \left(\frac{1+4.6(0.18)(0.932)}{1+6.1(0.18)}\right) = 0.844 \quad (2\text{-}4)$$

Wind pressure $$\frac{F}{A} = QK_zK_{zt}\left(V_{100}\right)^2 GC_f \quad (2\text{-}1a)$$

$$= (0.00256)(1.13)(1.0)(90)^2(0.844)(1.0)$$

$$= 19.8 \text{ psf}$$

Figure 5-1 shows tower areas and solidity ratios.

Transverse wind loads $C_f = 4.0\Phi^2 - 5.9\Phi + 4.0$

For $\Phi_{T1} = 0.69, C_{fT1} = 4.0(0.69)^2 - 5.9(0.69) + 4.0 = 1.83$, and $A_{T1} = 34 \text{ ft}^2$

For $\Phi_{T2} = 0.17, C_{fT2} = 4.0(0.17)^2 - 5.9(0.17) + 4.0 = 3.11$, and $A_{T2} = 33 \text{ ft}^2$

For $\Phi_{T3} = 0.15, C_{fT3} = 4.0(0.15)^2 - 5.9(0.15) + 4.0 = 3.21$, and $A_{T3} = 62 \text{ ft}^2$

where force coefficient equations are from Table 2-4

$$F_{T1} = 19.8(1.83)(34) = 1232 \text{ lb} = 1.2 \text{ kips}$$

$$F_{T2} = 19.8(3.11)(33) = 2032 \text{ lb} = 2.0 \text{ kips}$$

$$F_{T3} = 19.8(3.21)(62) = 3941 \text{ lb} = 3.9 \text{ kips}$$

5.0.3 Wind at 30°: Extreme Wind at 30° Yaw Angle (Chapter 2, Section 2.1)

Wind is at a 30° yaw angle.

5.0.3.1 Wind on Wires From the Wind load case, the wind pressure normal to the wires equals 16.1 psf

$$\text{Wind pressure} = 16.1\cos 2\,(\psi) = 16.1\cos 2(30) = 12.1 \text{ psf} \qquad (2\text{-}12)$$

Shield Wire Loads

$$V = 0.262(1800) + 50 = 522 \text{ lb} = 0.5 \text{ kips}$$

$$T = 12.1\left(\frac{0.385}{12}\right)(1500) + (2)(2472)\sin\left(\frac{5}{2}\right) = 798 \text{ lb} = 0.8 \text{ kips}$$

Conductor Loads

$$V = 1.075(1800) + 200 = 2135 \text{ lb} = 2.1 \text{ kips}$$

$$T = 12.1\left(\frac{1.165}{12}\right)(1500) + (2)(7396)\sin\left(\frac{5}{2}\right) = 2407 \text{ lb} = 2.4 \text{ kips}$$

Wind on Structure

From the Wind load case, the structure wind pressure equals 19.8 psf. Transverse force coefficients and areas are provided in the Wind loading case

$$\text{Wind force} = QK_zK_{zt}V_{\text{MRI}}^2G_t\left(1+0.2\sin^2(2\Psi)\right)\left(C_{ft}A_{mt}\cos^{2\cdot\cdot} + C_{fl}A_{ml}\sin^2\Psi\right) \qquad (2\text{-}14a)$$

$$C_f = 4.0\Phi^2 - 5.9\Phi + 4.0 \qquad \text{from Table 2-4}$$

For $\Phi_{L1} = 0.26, C_{fL1} = 4.0(0.26)^2 - 5.9(0.26) + 4.0 = 2.74$, and $A_{L1} = 43 \text{ ft}^2$

For $\Phi_{L2} = 0.24, C_{fL2} = 4.0(0.24)^2 - 5.9(0.24) + 4.0 = 2.81$, and $A_{L2} = 37 \text{ ft}^2$

For $\Phi_{L3} = 0.16, C_{fL3} = 4.0(0.16)^2 - 5.9(0.16) + 4.0 = 3.16$, and $A_{L3} = 67 \text{ ft}^2$

where force coefficient equations are from Table 2-4.

Transverse Wind Loads

$$\text{Transverse wind force} = 19.8\left(1+0.2\sin^2(2\Psi)\right)\times$$

$$\left(C_{ft}A_{mt}\cos^2\Psi + C_{fl}A_{ml}\sin^2\Psi\right)(\cos\Psi) \qquad (2\text{-}14a \text{ and } 2\text{-}14b)$$

$$F_{T1} = 19.8\left(1+0.2\sin^2\left(2(30)\right)\right)\left((1.83)(34)\cos^2(30)+(2.74)(43)\sin^2(30)\right)\left(\cos(30)\right) = 1501 \text{ lb} = 1.5 \text{ kips}$$

$$F_{T2} = 19.8\left(1+0.2\sin^2\left(2(30)\right)\right)\left((3.11)(33)\cos^2(30)+(2.81)(37)\sin^2(30)\right)\left(\cos(30)\right) = 2030 \text{ lb} = 2.0 \text{ kips}$$

$$F_{T3} = 19.8\left(1+0.2\sin^2\left(2(30)\right)\right)\left((3.21)(62)\cos^2(30)+(3.16)(67)\sin^2(30)\right)\left(\cos(30)\right) = 3987 \text{ lb} = 4.0 \text{ kips}$$

Longitudinal Wind Loads

$$\text{Longitudinal wind force} = 19.8\left(1+0.2\sin^2(2\Psi)\right)\left(C_{ft}A_{mt}\cos^2\Psi + C_{fl}A_{ml}\sin^2\Psi\right)\left(\sin\Psi\right)$$

(2-14a and 2-14c)

$$F_{L1} = 19.8\left(1+0.2\sin^2\left(2(30)\right)\right)\left((1.83)(34)\cos^2(30)+(2.74)(43)\sin^2(30)\right)\left(\sin(30)\right) = 867 = 0.9 \text{ kips}$$

$$F_{L2} = 19.8\left(1+0.2\sin^2\left(2(30)\right)\right)\left((3.11)(33)\cos^2(30)+(2.81)(37)\sin^2(30)\right)\left(\sin(30)\right) = 1172 = 1.2 \text{ kips}$$

$$F_{L3} = 19.8\left(1+0.2\sin^2\left(2(30)\right)\right)\left((3.21)(62)\cos^2(30)+(3.16)(67)\sin^2(30)\right)\left(\sin(30)\right) = 2302 = 2.3 \text{ kips}$$

5.0.4 Extreme Radial Glaze Ice with Wind (Chapter 2, Section 2.3)

Wind on Wires

From the Wind load case, K_z equals 1.21 and G_w equals 0.64.

From the wind and ice map [Figure 2-19(a)], t_{100} equals 0.25 in. and V_I equals 40 mph.

$$\text{Wind pressure} = QK_zK_{zt}\left(V_I\right)^2 G_wC_f \quad \text{(2-1a)}$$

$$= 0.00256(1.21)(1.0)(40)^2(0.64)(1.0)$$

$$= 3.17 \text{ psf}$$

$$t_z = t_{100}\left(\frac{z}{33}\right)^{0.10} = (0.25)\left(\frac{80}{33}\right)^{0.10} = 0.273 \text{ in.} \quad \text{(2-17a)}$$

Shield Wire Loads

$$W_i = 1.24(d+t_z)(t_z) \quad \text{(2-18)}$$

$$= 1.24(0.385+0.273)(0.273) = 0.223 \ \frac{\text{lb}}{\text{ft}}$$

$$d_i = 2(0.273)+0.385 = 0.931 \text{ in.}$$

$$V = (0.262 + 0.223)(1800) + 50 = 923 \text{ lb} = 0.9 \text{ kips}$$

$$T = 3.17\left(\frac{0.931}{12}\right)(1500) + (2)(2929)\sin\left(\frac{5}{2}\right) = 624 \text{ lb} = 0.6 \text{ kips}$$

Conductor Loads

$$W_i = 1.24(d + t_z)(t_z) \qquad (2\text{-}18)$$

$$= 1.24(1.165 + 0.273)(0.273) = 0.487 \ \frac{\text{lb}}{\text{ft}}$$

$$d_i = 2(0.273) + 1.165 = 1.711 \text{ in.}$$

$$V = (1.075 + 0.487)(1800) + 200 = 3012 \text{ lb} = 3.0 \text{ kips}$$

$$T = 3.17\left(\frac{1.711}{12}\right)(1500) + (2)(8092)\sin\left(\frac{5}{2}\right) = 1384 \text{ lb} = 1.4 \text{ kips}$$

Wind on Structure

From the Wind load case, K_z equals 1.13 and G_t equals 0.844.

$$\text{Wind pressure} = QK_zK_{zt}\left(V_{100}\right)^2 G_tC_f \qquad (2\text{-}1a)$$

$$= 0.00256(1.13)(1.0)(40)^2(0.844)(1.0)$$

$$= 3.91 \text{ psf}$$

Transverse Wind Loads

Force coefficients and areas are provided in the Wind loading case.

$$F_{T1} = 3.91(1.83)(34) = 243 \text{ lb} = 0.2 \text{ kips}$$

$$F_{T2} = 3.91(3.11)(33) = 401 \text{ lb} = 0.4 \text{ kips}$$

$$F_{T3} = 3.91(3.21)(62) = 778 \text{ lb} = 0.8 \text{ kips}$$

5.0.5 Construction and Maintenance (Chapter 3, Section 3.1)

Wind on Wires

Wind pressure = 3 psf

Shield Wire Loads

The pulling slope is 3 horizontal to 1 vertical

$$V = (1.5)(1757)\left(\frac{1}{3}\right) + (1.5)(0.262)\left(\frac{1800}{2}\right) + 1.5(50) = 1307 \text{ lb} = 1.3 \text{ kips}$$

(Alt. A controls)

$$V = (2)(0.262)(1800) + (2)(50) = 1043 \text{ lb} = 1.0 \text{ kips}$$

(Alt. B controls)

$$T = (1.5)(3)\left(\frac{0.385}{12}\right)(1500) + (1.5)(2)(1757)\sin\left(\frac{5}{2}\right) = 446 \text{ lb} = 0.4 \text{ kips}$$

Conductor Loads

$$V = (1.5)(5971)\left(\frac{1}{3}\right) + (1.5)(1.075)\left(\frac{1800}{2}\right) + 1.5(200) = 4737 \text{ lb} = 4.7 \text{ kips}$$

(Alt. A controls)

$$V = (2)(1.075)(1800) + (2)(200) = 4270 \text{ lb} = 4.3 \text{ kips}$$

(Alt. B controls)

$$T = (1.5)(3)\left(\frac{1.165}{12}\right)(1500) + (1.5)(2)(5971)\sin\left(\frac{5}{2}\right) = 1437 \text{ lb} = 1.4 \text{ kips}$$

Wind on Structure

Transverse Wind Loads

$$\text{Wind pressure} = (1.5)(3) = 4.5 \text{ psf}$$

Force coefficients and areas are provided in the Wind loading case.

$$F_{T1} = 4.5(1.83)(34) = 280 \text{ lb} = 0.3 \text{ kips}$$

$$F_{T2} = 4.5(3.11)(33) = 462 \text{ lb} = 0.5 \text{ kips}$$

$$F_{T3} = 4.5(3.21)(62) = 896 \text{ lb} = 0.9 \text{ kips}$$

5.0.6 Failure Containment (Chapter 3, Section 3.1.4 and Appendix I, Section 1.3.1)

This loading case is based on the residual static load of a broken conductor or shield wire (0 psf wind and 0 inches of radial ice at 30 °F).

Shield Wire Loads

The RSL load factor for a broken shield wire is 1.0.

Broken Wire

$$V = (0.262)\left(\frac{1800}{2}\right) + 50 = 286 \text{ lb} = 0.3 \text{ kips}$$

$$T = (1)(1628)\sin\left(\frac{5}{2}\right) = 71 \text{ lb} = 0.1 \text{ kips}$$

$$L = (1)(1628)\cos\left(\frac{5}{2}\right) = 1626 \text{ lb} = 1.6 \text{ kips}$$

Intact Wire

$$V = (0.262)(1800) + 50 = 522 \text{ lb} = 0.5 \text{ kips}$$

$$T = (2)(1628)\sin\left(\frac{5}{2}\right) = 142 \text{ lb} = 0.1 \text{ kips}$$

$$L = 0.0 \text{ kips}$$

Conductor Loads

$$\text{Ratio of the span to insulator length} = \frac{1250}{6} = 208$$

Note: For the purposes of this example, the span is taken to be the same as the Ruling Span.

$$\text{Ratio of the span to sag} = \frac{1250}{37.4} = 33$$

From Figure I-1, the RSL load factor is 0.7.

Broken Wire

$$V = (1.075)\left(\frac{1800}{2}\right) + 200 = 1168 \text{ lb} = 1.2 \text{ kips}$$

$$T = (0.7)(5622)\sin\left(\frac{5}{2}\right) = 172 \text{ lb} = 0.2 \text{ kips}$$

$$L = (0.7)(5622)\cos\left(\frac{5}{2}\right) = 3932 \text{ lb} = 3.9 \text{ kips}$$

Intact Wire

$$V = (1.075)(1800) + 200 = 2135 \text{ lb} = 2.1 \text{ kips}$$

$$T = (2)(5622)\sin\left(\frac{5}{2}\right) = 490 \text{ lb} = 0.5 \text{ kips}$$

$$L = 0.0 \text{ kips}$$

5.1 WEIGHT SPAN CHANGE WITH BLOWOUT ON INCLINED SPANS

This example compares weight spans with and without wind for the center tower shown in Figure 5-2. The equations are shown in Section 4.5.3 Chapter 4. Wire data are from Section 5.

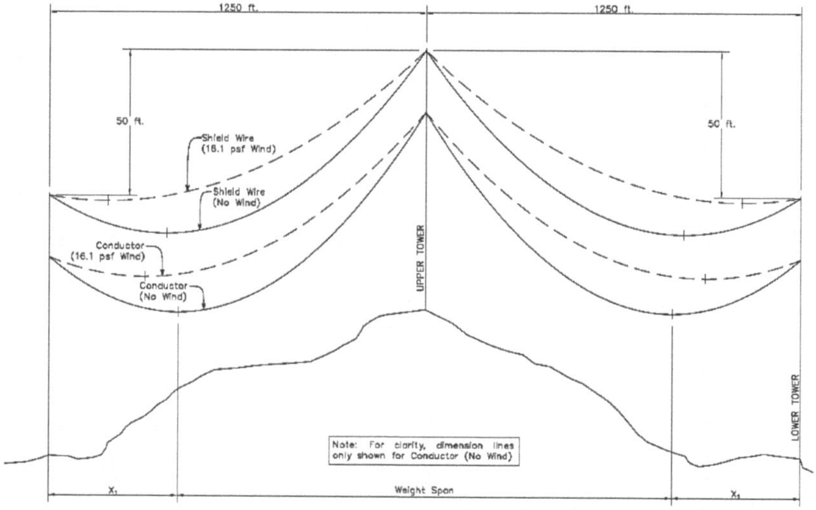

Figure 5-2. Weight span for center tower with inclined spans.

Shield Wire

No Wind

$$C_v = \frac{H}{w_v} = \frac{1550}{0.262} = 5916 \text{ ft} \qquad \text{from Section 4.5.2}$$

$$X_1 = \frac{S}{2} - C_v \sinh^{-1}\left(\frac{\frac{B}{2}}{C_v \sinh\left(\frac{S}{2C_v}\right)}\right)$$

$$X_1 = \frac{1250}{2} - 5916 \sinh^{-1}\left(\frac{\frac{50}{2}}{5916 \sinh\left(\frac{1250}{2(5916)}\right)}\right) = 389 \text{ ft} \qquad (4\text{-}5)$$

$$\text{Weight Span} = 2(1250 - 389) = 1722 \text{ ft}$$

16.1 psf Wind

$$C_v = \frac{H}{w_v} = \frac{2917}{0.262} = 11134 \text{ ft} \qquad \text{from Section 4.5.2}$$

$$X_1 = \frac{S}{2} - C_v \sinh^{-1}\left(\frac{\frac{B}{2}}{C_v \sinh\left(\frac{S}{2C_v}\right)}\right)$$

$$X_1 = \frac{1250}{2} - 11134 \sinh^{-1}\left(\frac{\frac{50}{2}}{11134 \sinh\left(\frac{1250}{2(11134)}\right)}\right) = 180 \text{ ft} \qquad (4\text{-}5)$$

$$\text{Weight Span} = 2(1250 - 180) = 2140 \text{ ft} (24\% \text{ increase})$$

Conductor

No Wind

$$C_v = \frac{H}{w_v} = \frac{5305}{1.075} = 4935 \text{ ft} \qquad \text{from Section 4.5.2}$$

$$X_1 = \frac{S}{2} - C_v \sinh^{-1}\left(\frac{\frac{B}{2}}{C_v \sinh\left(\frac{S}{2C_v}\right)}\right) \tag{4-5}$$

$$X_1 = \frac{1250}{2} - 4935 \sinh^{-1}\left(\frac{\frac{50}{2}}{4935 \sinh\left(\frac{1250}{2(4935)}\right)}\right) = 428 \text{ ft}$$

$$\text{Weight Span} = 2(1250 - 428) = 1644 \text{ ft}$$

16.1 psf Wind

$$C_v = \frac{H}{w_v} = \frac{8530}{1.075} = 7935 \text{ ft} \qquad \text{from Section 4.5.2}$$

$$X_1 = \frac{S}{2} - C_v \sinh^{-1}\left(\frac{\frac{B}{2}}{C_v \sinh\left(\frac{S}{2C_v}\right)}\right)$$

$$X_1 = \frac{1250}{2} - 7935 \sinh^{-1}\left(\frac{\frac{50}{2}}{7935 \sinh\left(\frac{1250}{2(7935)}\right)}\right) = 308 \text{ ft} \tag{4-5}$$

$$\text{Weight Span} = 2(1250 - 308) = 1884 \text{ ft(15\% increase)}$$

5.2 TRADITIONAL CATENARY CONSTANT

This example compares weight spans to those in Section 5.1 using the traditional catenary constant. The traditional catenary constant is based on the resultant unit weight (w_r). The catenary constant in Section 5.1 is based on the vertical unit weight (w_v). Figure 5-2 shows the upper and lower towers and spans.

Shield Wire

16.1 psf Wind

$$w_v = 0.262 \ \frac{\text{lb}}{\text{ft}}$$

$$w_t = 16.1\left(\frac{0.385}{12}\right) = 0.517 \ \frac{\text{lb}}{\text{ft}}$$

$$w_r = \sqrt{(0.262)^2 + (0.517)^2} = 0.580 \ \frac{\text{lb}}{\text{ft}}$$

$$C_r = \frac{T_H}{w_r} = \frac{2917}{0.580} = 5029 \text{ ft} \tag{4-8}$$

$$X_1 = \frac{S}{2} - C_r \sinh^{-1}\left(\frac{\frac{B}{2}}{C_r \sinh\left(\frac{S}{2C_r}\right)}\right) \tag{4-5}$$

$$X_1 = \frac{1250}{2} - 5029\sinh^{-1}\left(\frac{\frac{50}{2}}{5029\sinh\left(\frac{1250}{2(5029)}\right)}\right) = 424 \text{ ft}$$

$$\text{Weight Span} = 2(1250 - 424) = 1652 \text{ ft}$$

Conductor

Refer to Table 5-3.

16.1 psf Wind

$$w_v = 1.075 \ \frac{\text{lb}}{\text{ft}}$$

$$w_t = 16.1\left(\frac{1.165}{12}\right) = 1.563 \ \frac{\text{lb}}{\text{ft}}$$

$$w_r = \sqrt{(1.075)^2 + (1.563)^2} = 1.897 \ \frac{\text{lb}}{\text{ft}}$$

$$C_r = \frac{T_H}{w_r} = \frac{8530}{1.897} = 4497 \text{ ft} \quad \text{(4-8)}$$

$$X_1 = \frac{S}{2} - C_r \sinh^{-1}\left(\frac{\frac{B}{2}}{C_r \sinh\left(\frac{S}{2C_r}\right)}\right) \quad \text{(4-5)}$$

$$X_1 = \frac{1250}{2} - 4497 \sinh^{-1}\left(\frac{\frac{50}{2}}{4497 \sinh\left(\frac{1250}{2(4497)}\right)}\right)$$

$$= 446 \text{ ft}$$

$$\text{Weight Span} = 2(1250 - 446) = 1608 \text{ ft}$$

The traditional catenary constant underestimates the vertical load on the upper tower and overestimates the vertical load on the lower tower.

APPENDIX A

DEFINITIONS, NOTATIONS, AND SI CONVERSION FACTORS

A.1 GENERAL DEFINITIONS

Conductor Creep: Permanent elongation of conductors under everyday tension conditions (Aluminum Company of America 1961).

MRI: Mean Recurrence Interval (years), also known as "Return Period." The inverse of the probability of exceedance of an environmental load (i.e., wind, ice) in any given year. For example, a design event with the probability of exceedance of 0.01 (1%) is associated with an MRI of 100 years.

Longitudinal: Local axis of structure that is, for tangent structures, parallel to the direction of the conductors and perpendicular to the vertical axis. In angle structures, this is generally the direction perpendicular to the angle bisector.

Transverse: Local axis of structure that is, for tangent structures, perpendicular to the direction of the conductors and to the vertical axis. In angle structures, this is the direction of the angle bisector.

Vertical: Local axis of structure that is parallel with the direction of gravitation force.

A.2 DEFINITIONS OF STRUCTURE TYPES

Tangent Structures: Structures having minimum line deflection angle, typically less than 2°, such that transverse loads resulting from conductor tension are relatively small compared to those resulting from other sources. Structures which primarily resist the vertical weight of the wires with any accumulated ice and the transverse loads due to wind.

Tangent structures typically utilize suspension conductor connections; however, other means of attachment, such as post insulators, are common.

Angle Structures: Structures marking changes in line angle along the transmission line which support the same vertical and transverse loads as tangent structures, as well as transverse loads resulting from wire tensions being applied at an angle.

- May be similar to tangent structures, using suspension or post insulators to support the conductors and transfer wind, weight, and line angle loads to the structure.
- May be similar to strain or dead-end structures, using insulators in series with the conductors to bring wind, weight, and line angle loads directly into the structure.

Dead-End Structures: Structures designed as termination points for wires and capable of supporting the loads resulting from the removal of all wires from one or more spans.

Strain Structures: Structures similar to dead-end structures, where wire tensions are transferred directly to the structure but are capable of resisting only unbalanced/differential tensions.

A.3 DEFINITIONS OF SPAN

Span: Unless otherwise stated, span usually refers to the distance between two adjacent structures, generally measured horizontally (Figure A-1).

Ahead Span: The span in front (generally in the direction of increasing stationing or ascending structure numbering) of the structure in question. In Figure A-1, Span 2 is the ahead span of Structure 11.

Back Span: The span behind (generally in the direction of decreasing stationing or descending structure numbering) the structure in question. In Figure A-1, Span 1 is the back span of Structure 11.

Sag: The distance measured vertically from a conductor to the straight line joining its two points of support.

Slack: The amount of conductor length difference between a straight line made by two adjacent supports and a sagging conductor.

Weight Span: The horizontal distance between the low point of sag of adjacent spans. It is used in calculating the vertical load the conductor imposes on the supporting structure (Figure A-1). This may also be referred to as the vertical span.

Wind Span: The mathematical average of the back span and the ahead span. It is used in calculating the wind load the conductor imposes on the supporting structure. This may also be referred to as the horizontal span or the transverse span.

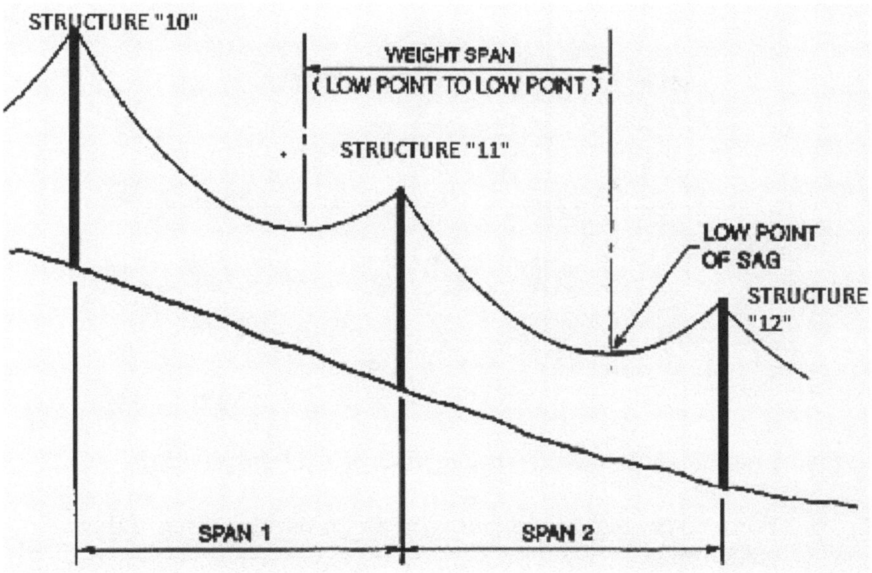

Figure A-1. Span usually refers to the distance between two adjacent structures.

A.4 NOTATION

Unless otherwise stated, the following notation is used in this manual:

A	Solid tributary area of surfaces projected normal to the wind
A_{ml}	Area of all members on the longitudinal face of the structure (Figure A-2)
A_{mt}	Area of all members on the transverse face of the structure (Figure A-2)
A_o	Area of the outline of the windward face of the structure
A_s	Projected area of marker ball
B_t	Dimensionless response term corresponding to the quasi-static background wind loading on the structure
B_w	Dimensionless response term corresponding to the quasi-static background wind loading on the wire (conductor or shield wire)
BWL	Broken wire load
C_{exp}	Turbulence intensity constant
C_f	Force coefficient associated with the windward face of a component
C_{fl}	Force coefficient associated with the longitudinal faces of the structure

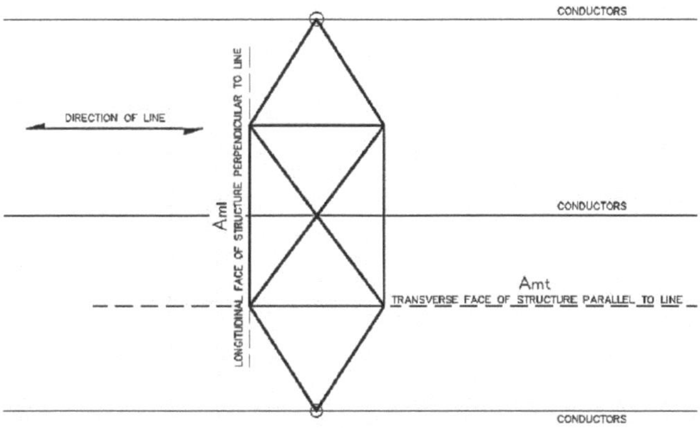

Figure A-2. Plan view: Structure longitudinal and transverse axis

C_{ft}	Force coefficient associated with the transverse faces of the structure
C&M	Construction and maintenance loads
COV_R	Coefficient of variation of component strength
d	Diameter of wire (conductor or shield wire)
d_s	Projected diameter of wire (conductor or shield wire) with ice accretion as appropriate
D_i	Straight-line distance between the supports
D_j	Practical jet diameter
E	Modulus of elasticity
EDT	Everyday wire tension
F	Wind force
f	Structure/member natural frequency
F_d	Force in direction of wind
F_l	Wind force in the longitudinal direction
F_t	Wind force in the transverse direction
f_t	Fundamental frequency of the free-standing structure in the transverse direction
f_w	Fundamental frequency for horizontal sway of the conductor or shield wire
FC	Failure containment loads
G	Gust response factor
G_t	Gust response factor for the structure
G_w	Gust response factor for the wire (conductor or shield wire)
H	Height of hill or escarpment relative to the upwind terrain
h	Insulator length
HA	Horizontal line angle

HIW	High intensity wind
I_z	Turbulence intensity at effective height of structure
K_z	Wind pressure exposure coefficient, also known as terrain factor at height z aboveground
K_{zt}	Topographic factor
K_1	Factor to account for shape of topographic feature and maximum speed-up effect
K_2	Factor to account for reduction in speed-up with distance upwind of downwind effect
K_3	Factor to account for reduction in speed-up with height above local terrain
L	Unit length of conductor
l	Length of wire from low point to support
L_h	Distance upwind of crest to where the difference in ground elevation is half the height of hill or escarpment
L_m	Length of member
L_s	Transverse integral scale of turbulence
L_T	Transverse structure load
L_V	Vertical structure load
LC	Load case
m_i	Unit mass of typical ice sample
N	Period of time
PDF	Probability density function of a random variable
Q	Air density coefficient
Q_i	Numerical constant to convert radial ice thickness to weight
Q_{MRI}	Reliability adjustment factor
R	Radial polar coordinate of downburst
r	Radius of an aerial marker ball
R_e	Reynolds number
R_n	Nominal value of component strength
RSL	Residual static load for the broken wire loading condition
R_t	Dimensionless resonant response term of the structure
R_w	Dimensionless resonant response term of the wire
R_X	Wire force in the longitudinal direction
R_Y	Wire force in the transverse direction
S	Span length of the wires (conductor and ground wire)
s	Member diameter or width normal to the wind
S_t	Strouhal number
T	Transverse load due to wind
t	Nominal ice thickness
T_H	Horizontal component of tension
t_z	Design ice thickness for heights z above ground
V_c	Velocity of concurrent wind speed with ice
V_{cr}	Critical vortex-induced wind speed

V_{eqc}	Equivalent uniform velocity profile for wires
V_{eqt}	Equivalent uniform velocity profile for structures
V_i	Volume of ice accreted on aerial marker ball
V_J	Velocity of dropping cold air jet in a downburst model
V_{mean}	Mean hourly wind speed
V_{MRI}	Basic wind speed for select MRI, 3-second gust at 33 ft (10 m) height in open country
$\overline{V}_{\text{MRI}}$	Basic wind speed, converted to hourly average
V_{RD}	Downburst outflow velocity radial component
V_{VR}	Downburst outflow velocity vertical component
V_{100}	Basic wind speed at a 100-year mean recurrence interval, 3-second gust
VA	Vertical line angle
w	Wire weight per unit length
w_c	Unit weight of wire
W_G	Tornado gust width
w_h	Unit transverse wire load
W_i	Weight of glaze ice
W_p	Wind pressure
w_r	Resultant unit weight
w_t	Unit weight, wind component
w_v	Vertical unit weight
X_L	Distance from low point of sag to lower support
X_R	Distance from low point of sag to upper support
z	Height above ground
z_g	Gradient height
z_h	Effective height above ground of the wire (conductor or ground wire) or structure
α	Power law exponent for gust wind
$\overline{\alpha}$	Power law exponent for mean hourly wind
γ	Load factor applied to weather-related loads
ε	Coefficient for separation of the wire and structure response terms in the general gust response factor equations
ζ_t	Ratio of calculated structure damping to critical structure damping
ζ_w	Ratio of calculated wire damping to critical wire damping
κ	Surface drag coefficient
ρ	Mass density of air
ρ_i	Ice density
Φ	Solidity ratio (Am/Ao) or strength reduction factor
Ψ	Angle of yaw

A.4 SI CONVERSION FACTORS

1 ft = 0.3048 meter (m)
1 in. = 25.4 millimeters (mm)
1 pound (lb) force = 4.45 Newtons (N)
1 lb/ft = 14.6 N/m
1 lb/ft^2 (psf) = 47.8 Pascals (Pa) (N/m^2)
1 pcf = 0.016 gram/cubic centimeter (g/cm^3)
1 mile per hour (mph) = 0.45 meter/second (m/s)

APPENDIX B
RELIABILITY-BASED DESIGN

Additional information on Reliability-Based Design (RBD) methodology, as applicable to transmission line structures, can be found in ASCE Manual of Practice No. 111, *Reliability-Based Design of Utility Pole Structures* (2006).

APPENDIX C
AIR DENSITY COEFFICIENT, Q

The air density coefficient, Q, converts the kinetic energy of moving air into potential energy of pressure. The value of Q can be determined from Equation (C-1)

$$Q = 0.5\rho \tag{C-1}$$

where ρ is the mass density of air.

The standard value of Q based on the specific weight of air at 59 °F (15 °C) at sea level pressure of 14.7 psi (101.325 kPa) is 0.00256 for use with customary units (0.613 for use with SI units). For customary units, the dimensions of Q are associated with wind speed in miles per hour and pressure in pounds per square foot. For SI units, the dimensions of Q are associated with wind speed in meters per second and pressure in pascals. The use of any other value for Q for a nonstandard temperature or elevation should be based on sound engineering judgment with sufficient meteorological data available to justify a different value for a specific design application.

The specific weight of air varies with temperature and atmospheric pressure. Table C-1 shows values of the air density as a function of air temperature and elevation above sea level. The effect of moisture or variation in relative humidity is assumed to be negligible.

Table C-1. Air Density Coefficient, *Q*

Air temperature (°F)	Elevation above sea level (ft)					
	0	2,000	4,000	6,000	8,000	10,000
100	0.00238	0.00221	0.00205	0.00191	0.00177	0.00165
80	0.00246	0.00229	0.00213	0.00198	0.00184	0.00171
60	0.00256*	0.00237	0.00221	0.00205	0.00191	0.00178
40	0.00266	0.00247	0.00230	0.00214	0.00199	0.00185
20	0.00277	0.00257	0.00239	0.00223	0.00207	0.00192
0	0.00289	0.00268	0.00249	0.00232	0.00216	0.00201
–20	0.00293	0.00281	0.00261	0.00243	0.00226	0.00210
–40	0.00317	0.00294	0.00273	0.00254	0.00237	0.00220

Source: Adapted from Brekke (1959).

* Recommended value

APPENDIX D
CONVERSION OF WIND SPEED AVERAGING TIME

It is recognized that wind speed values for a given record depend on the averaging time used in the measurement of the wind speed statistics. The use of a shorter averaging time results in a higher wind speed, whereas a longer averaging time results in a lower wind speed. This is due to the natural gusts and calms in wind patterns. It is often necessary to obtain equivalent wind speeds based on different averaging periods. Conversion of a wind speed to that representative of another averaging time can be accomplished using the relationship shown in Figure D-1. This graph, prepared from results by Durst (1960), gives the ratio, $(V_t/V_{1\text{-hour}})$, of probable maximum wind speed averaged over $_t$ seconds to hourly mean wind speed for Exposure Category C. Note that in the graph, the value V_{3600} represents the wind speed averaged over 3,600 seconds (1 hour). Additional discussion of the Durst gust factor curve can be found in Miller (2011).

The hurricane simulation technique (ASCE 2017) used to develop the wind speed maps referred to in this manual apply the relationship of gust wind speeds described in Engineering Science Data Unit (ESDU) (1982, 1983), which have been validated for hurricane winds by Vickery and Skerlj (2005) and Jung and Masters (2013).

The calculation of the resonant component of the dynamic response for the gust response factor (see Appendix F) requires the mean hourly wind speed to be calculated from the 3-second gust wind speed. The following is an example of converting a 3-second gust wind speed (such as those specified in Figure 2-1) to a mean hourly wind speed. This relationship is also given in Equation (F-18) in Appendix F.

- Step 1. Obtain $V_{3\text{sec}}$ from Figure 2-1. This is the wind speed averaged more than 3 seconds at a height of 33 ft (10 m) in open country terrain.

- Step 2. Obtain the ratio ($V_t/V_{1\text{-hour}}$) for t = 3 seconds from Figure D-1. This ratio is commonly taken to be 1.52.
- Step 3. Calculate the mean hourly wind speed, V_{mean}, based on $V_{mean} = \frac{V_{3sec}}{1.52}$.

For example, if Figure 2-1 indicated that a 3-second gust wind speed of 90 mph was to be used, the corresponding mean hourly wind speed is approximately 59 mph. Note that the example provided is specific for conversion from a 3-second gust, although the conversion to other averaging times is carried out in a similar fashion using other values from Figure D-1.

Additional relationships (e.g., IEC 2003) have been developed to address conversion of wind speed averaging time, and often yield varying results. This is particularly true for Exposures B and D, where only limited historical data are available for extreme wind events. Due to the uncertainties related to wind and its measurement, it is recommended that the conversion of wind speed averaging time should only be applied to wind speeds associated with open country terrain exposure (Exposure Category C).

If conversions of wind speed averaging times based on other exposures are of interest to the user, the IEC 60826 (IEC 2003a, b) provides different conversion factors based on terrain exposure categories. Alternate values may be used for conversion of wind speed averaging time for Exposures B or D if reasonable values can be determined from the use of local wind data.

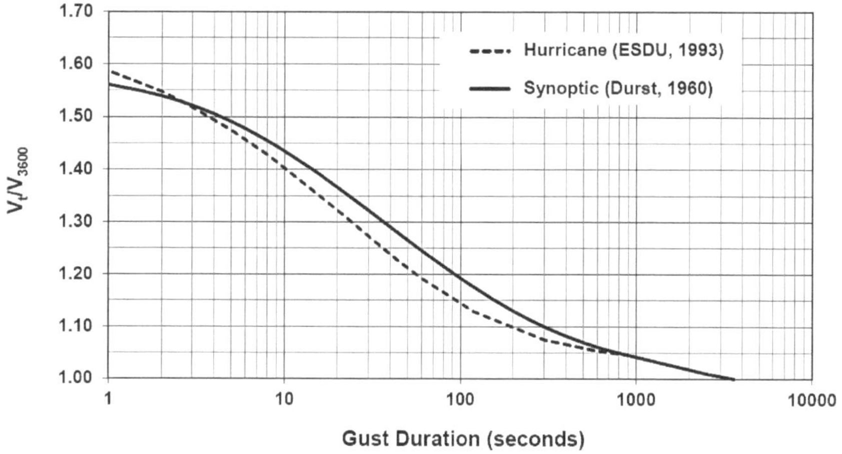

Figure D-1. Conversion relationship for wind speed averaging time, Exposure C. Source: ASCE (2017).

APPENDIX E
SUPPLEMENTAL INFORMATION ON STRUCTURE VIBRATION

E.1 INTRODUCTION

A transmission line has a large number of structures located at sites with varying environmental and geographical exposures. Given the long length of transmission lines, the probability of transmission structures or their components being located in an environment prone to inducing vibration is greater than that for non-transmission line structure types.

Vibration of a transmission structure can consist of complete structure vibration modes, structure component vibration modes, or individual member vibration modes. The initiation of these modes can be caused by vibration forces induced either by the wind acting directly on the structure or the conductor and overhead ground wires. Although general precautions during the initial design of transmission structures can be considered to reduce the possibility of vortex-induced vibration problems (ASCE 1961), the occurrence of vibration problems is highly dependent on steady-state wind, terrain, orientation of the wind to the member cross section, and local conditions.

E.2 STRUCTURE VIBRATION DUE TO WIND ON STRUCTURE

The majority of observed cases of member vibration can be attributed to the vortex shedding phenomenon (e.g., von Kármán vortex street). In this phenomenon, vortices are shed from the member in an alternating periodic pattern, which results in oscillating forces on the member in the plane perpendicular to the wind. If a natural frequency of the member aligns with the frequency associated with vortex shedding, then a condition of harmonic resonance can occur. This is commonly referred to as vortex-induced vibration (VIV), and may occur for a single member or the entire structure.

Vortex shedding is related to the shape and size of the member, the frequency of the member, wind speed, and wind orientation. The shape of the member is taken into account through the use of the Strouhal number, St, which is a nondimensional parameter. Based on the frequency of the member, the corresponding Strouhal number, and the dimension of the member exposed to the wind, a wind speed at which vortex shedding can occur for the section can be calculated. This is referred to as the critical wind speed and is shown in Equation (E-1):

$$V_{cr} = \frac{f\,s}{\mathrm{St}} \quad \text{(E-1)}$$

where

V_{cr} = Critical wind speed associated with vortex sheading [ft/s (m/s)],
f = Structure or member natural frequency (Hertz),
St = Strouhal number, and
s = Across-wind dimension [ft (m)].

Standard structural shapes have an average Strouhal number of 0.14. Strouhal numbers for a variety of structural shapes can be found in Simiu and Scanlan (1996). The structure's natural frequency can be determined using structural dynamic theory (Clough and Penzien 1975, Mathur et al. 1986, Paz 1980, Trainor et al. 1984). Vortex-induced motion can cause flexural, torsional, or coupled flexural-torsional vibration modes.

Tubular steel pole arms and/or crossarms may be susceptible to damage from vortex-induced vibration. The symmetrical shape of the arms and the absence of the vibration damping effect of attached conductors, ground wires, and assemblies may result in structural damage in the form of fatigue cracking and complete failure, most commonly in relatively low wind velocities. Cyclical stresses at the support due to vortex shedding of the arm can exceed the endurance limits in a relatively short time period. This cyclic loading can lead to cracking and premature failure. When stringing operations do not occur soon after installation of tubular arms, the installation of a weight at the end of the arm or tying the end of the arm to the pole are common practices to reduce possible vortex-induced vibration. ASCE 48-11 (2011) contains additional information on vibration of unloaded arms and common remedial measures to dampen oscillations.

A latticed tower, in general, presents a complex aerodynamic shape to the wind such that consistent vortex shedding causing complete oscillation of the structure over a prolonged period is unlikely. Therefore, only individual member behavior of latticed towers subjected to vortex shedding and aeroelastic instability has been studied (Modi and Slater 1983, Wardlaw 1967). Structural shapes (Thrasher 1984) and cable components, such as guy wires and cable tension members, can be excited by vortex-induced vibration. Latticed tower members, which are long and flexible, are particularly susceptible to vortex-induced vibration.

Large latticed structures for heavy angle sites and river crossings frequently make use of stitched double-angle members. These members have been found to be very susceptible to torsional flutter in moderate winds. For these double angles, slenderness ratios should not exceed 200. Vibration can be reduced using wind spoilers. One example of a wind spoiler for this application is the insertion of small, flat plates that project beyond the horizontal legs of the angle, commonly referred to as splitter plates.

Utilities have experienced failures of the end connection plates or coped connecting members from this wind-induced action. This phenomenon has been documented by wind tunnel tests.

Solutions to vortex-induced problems developing during the service life of the tower can consist of changing the frequency of the member, increasing the damping, or by changing the aerodynamic properties of the member. Changing the member cross section, or member boundary conditions (connections), or adding intermediate structural bracing can modify the member frequency.

Although infrequent in transmission structures, aeroelastic instability of certain structural shapes can be a potential problem. Wind forces acting on a structural shape that is inherently unstable at certain wind angles cause this wind-induced vibration. Additional information can be found in Houghton and Carruthers (1976), MacDonald (1975), Modi and Slater (1983), Sachs (1972), Simiu and Scanlan (1996), and Slater and Modi (1971).

E.3 STRUCTURE VIBRATION DUE TO WIRE MOTION

Structure or member vibration can occur from conductor motion (Aeolian, sub-conductor oscillation, and galloping) when the frequency of the vibrating conductor corresponds to one of the natural frequencies of the structure or its individual member(s). Approximate natural frequencies of conductor vibration for Aeolian motion are 3 to 150 Hertz; for sub-conductor oscillation, 0.15 to 10 Hertz; and for galloping, 0.08 to 3 Hertz (EPRI 1979). In most instances conductor systems can be designed using dampers and spacer dampers to prevent and/or reduce the effect of wind-induced vibration behavior. Although Aeolian vibration of wires has caused some instances of fatigue failures of structure or hardware elements, galloping wires have the potential to cause the most damage to the supporting structure (Brokenshire 1979, Gibbon 1984, White 1979). Latticed steel running angle suspension towers, guyed-mast dead-end structures, heavy angle towers, and flexible "narrow-base" pole structures have been reported to be more susceptible to damage caused by conductor galloping motion. Field investigations have been conducted to study methods of suppressing conductor galloping (Pohlman and Havard 1979, Richardson 1983).

APPENDIX F
EQUATIONS FOR GUST RESPONSE FACTORS

F.1 INTRODUCTION

The gust response factor (GRF) accounts for the load effects due to wind turbulence and dynamic amplification of flexible structures and cables. It represents the cumulative effect of the time-varying and spatially varying fluctuating wind speeds over the range of span lengths of typical transmission lines, as well as the effect of the wind on supporting structures. The approach for the gust response factors provided in Section 2.1.5 of Chapter 2 of this manual are based on work by Davenport (1979) for estimating the peak response of transmission line systems to gusting winds, as well as wind loading provisions in ASCE 7-16 (2017).

The original Davenport GRF equations were developed using statistical methods which involve the spatial correlation and energy spectrum of turbulent wind, as well as the dynamic characteristics of transmission line components. The complete GRF equations include amplification factors that account for the resonant component of the dynamic response of structures and wires. The derivation of the gust response factor is given in Davenport (1979), and their application to typical towers and wires is discussed in previous versions of this manual (ASCE 1991, 2010a). In the second edition of this manual (ASCE 1991), simplified equations are presented where the resonant component of the dynamic response is negligible for a large range of typical tower configurations. These simplifications were based on a theoretical appraisal of transmission line behavior, as well as an assessment of available full-scale data. The underlying assumptions and limitations of the simplified procedure are discussed further in Section F.4. In the third edition of this manual (ASCE 2010a), the gust response factors were modified to be made compatible with the 3-second gust wind speed for consistency with ASCE 7.

The equations for some of the components of the gust response factor have been modified for this edition of the manual. Most notable are the removal of the parameters κ (surface drag coefficient) and E (exposure factor); these have been replaced with more current parameters used in the description of atmospheric boundary layer wind. As well, the exponents for the power law now reflect a mean hourly wind speed for which the equations were derived. This is in contrast to the use of power law exponents for the fastest-mile wind speed used previously. The updated approach involves the calculation of the turbulence intensity of the wind at the effective height of the structure or wires, as well as the use of separate peak factors for the background and resonant components of the dynamic response. The revisions to the methodology reflect the state of the art in the calculation of wind loads on structures and are consistent with the methodology applied in ASCE 7. A detailed description of these changes is given by Mara (2015).

These equations are based on idealized conditions that may or may not reflect the true weather events that a transmission line structure may experience. Thus, the results obtained by the application of these equations within this context should be considered approximate. The purpose of this appendix is to present the gust response factor equations and define the various wind, exposure, and dynamic parameters used. The approach for the gust response factor of structures and lines is consistent with that developed by Davenport (1979); however, some of the nomenclature has been slightly modified to incorporate the relationships used in the development of the ASCE 7-16 wind load criteria that form the basis of this manual.

The equations are given in this appendix without derivation. However, interested readers may refer to several papers that have dealt with this subject (e.g., Davenport 1962, 1967, 1977, 1979; Vellozzi and Cohen 1968).

F.2 NOMENCLATURE

The following nomenclature is used in this appendix:

B_t Dimensionless response term corresponding to the quasi-static background component of the dynamic response of the tower [Equation (F-6)]

B_w Dimensionless response term corresponding to the quasi-static background component of the dynamic response of the wires [Equation (F-14)]

c_{exp} Turbulence intensity constant, based on exposure (Table F-1)
C_f Force coefficient for the wires (typically taken as 1.0; see Appendix G)
d_s Diameter of the wire (conductor or shield wire) in inches
f_t Fundamental frequency of the tower or structure in the transverse direction, in Hz (see Table F-2 for approximate values)
f_w Fundamental frequency for horizontal sway of the conductor or shield wire, in Hz [Equation (F-20)]
g_B Peak factor for the background component of the dynamic response (same for tower and wires, constant value of 3.6)
g_{Rt} Peak factor for the resonant component of the dynamic response of the tower [Equation (F-4)]
g_{Rw} Peak factor for the resonant component of the dynamic response of the wires [Equation (F-12)]
g_v Peak factor for the turbulence of the wind (constant value of 3.6)
G_t Gust response factor for wind loading on structure [Equation (F-1)]
G_w Gust response factor for wind loading on wires [Equation (F-9)]
I_z Turbulence intensity at effective height of the tower/structure or wire [Equations (F-2) and (F-10)]
L_s Integral length scale of turbulence (ft) (Table F-1)
R_t Dimensionless resonant response term of the structure [Equation (F-7)]
R_w Dimensionless resonant response term of the wires [Equation (F-15)]
S Design wind span (ft)
Sag Wire sag at midspan (ft)
V_{MRI} Basic wind speed, 3-second gust at 10 m height in open country terrain (Figure 2-1)
$\overline{V}_{MRI}$ Basic wind speed, converted to mean hourly wind speed (Appendix D)
V_o Mean hourly wind speed (ft/s) at effective height of the tower/structure or wires, based on exposure
z_g Gradient height (ft) (Table F-1)
z_h Effective height of wires and/or structure (Section 2.1.4.3)
$\overline{\alpha}$ Power law exponent for mean hourly wind (Table F-1)
α Power law exponent for gust wind (Table F-1)
ε Separation coefficient reflecting the non-coincidence of tower and wire loads (for typical transmission line systems, ε is approximated by 0.75)
ζ_t Damping ratio of structure relative to critical (see Table F-2 for approximate values)
ζ_w Damping ratio of wires relative to critical [Equation (F-21)]

F.3 RELEVANT EQUATIONS

The gust response factors for the tower, G_t, and the wires, G_w, are given in Equations (F-1) and (F-9), respectively. All parameters are described in Section F.2, and wind parameters by exposure category are listed in Table F-1. Note that the subsequent equations contain both the background and resonant components of the dynamic response, as opposed to the simplified versions provided in Section 2.1.5.1. For unique structures (e.g., complex structural configurations, very tall structures, structures with low fundamental frequencies) and for very long spans, the inclusion of the resonant component may be of importance. As the magnitude of the resonant effect decreases, Equations (F-1) and (F-9) converge to Equations (F-8) and (F-16), respectively.

Table F-1. Wind Parameters by Exposure Category

Exposure	α	$\bar{\alpha}$	c_{exp}	L_s (ft)	z_g (ft)
B	7.0	4	0.3	170	1200
C	9.5	6.5	0.2	220	900
D	11.5	9	0.15	250	700

Source: Adapted from ASCE 7-16 (ASCE 2017).

The gust response factor for the tower, G_t, with an effective height z_h is calculated as

$$G_t = \left(\frac{1 + 1.7 I_z \varepsilon \sqrt{g_B^2 B_t^2 + g_{Rt}^2 R_t^2}}{1 + 1.7 I_z g_v} \right) \tag{F-1}$$

in which

$$I_z = c_{exp} \left(\frac{33}{z_h} \right)^{\frac{1}{6}} \tag{F-2}$$

$$g_b = g_v = 3.6 \tag{F-3}$$

$$g_{Rt} = \sqrt{2\ln(3600 f_t)} + \frac{0.577}{\sqrt{2\ln(3600 f_t)}} \tag{F-4}$$

$$\varepsilon = 0.75 \tag{F-5}$$

$$B_t = \sqrt{\frac{1}{1+\frac{0.56z_h}{L_s}}} \tag{F-6}$$

$$R_t = \sqrt{\left(\frac{0.0123}{\zeta_t}\right)\left(\frac{f_t z_h}{V_o}\right)^{\frac{-5}{3}}} \tag{F-7}$$

Note that if the resonant component is neglected, Equation (F-1) converges to the simplified expression presented in Chapter 2:

$$G_t = \left(\frac{1+4.6I_z B_t}{1+6.1I_z}\right) \tag{F-8}$$

The gust response factor for the wires, G_w, with an effective height z_h is calculated as

$$G_w = \left(\frac{1+1.7I_z\varepsilon\sqrt{g_B^2 B_w^2 + g_{Rw}^2 R_w^2}}{1+1.7I_z g_v}\right) \tag{F-9}$$

in which

$$I_z = c_{\text{exp}}\left(\frac{33}{z_h}\right)^{\frac{1}{6}} \tag{F-10}$$

$$g_b = g_v = 3.6 \tag{F-11}$$

$$g_{Rw} = \sqrt{2\ln(3600 f_w)} + \frac{0.577}{\sqrt{2\ln(3600 f_w)}} \tag{F-12}$$

$$\varepsilon = 0.75 \tag{F-13}$$

$$B_w = \sqrt{\frac{1}{1+\frac{0.8S}{L_s}}} \tag{F-14}$$

$$R_w = \sqrt{\left(\frac{0.0113}{\zeta_w}\right)\left(\frac{z_h}{S}\right)\left(\frac{f_w z_h}{V_o}\right)^{\frac{-5}{3}}} \tag{F-15}$$

Note that if the resonant component is neglected, Equation (F-9) converges to the simplified expression presented in Chapter 2:

$$G_w = \left(\frac{1+4.6 I_z B_w}{1+6.1 I_z}\right) \tag{F-16}$$

For the calculation of R_t and R_w, the resonant components of the dynamic response, the mean hourly wind speed (ft/s) at the effective tower and wire heights is calculated as

$$V_o = 1.66\left(\frac{z_h}{z_g}\right)^{\frac{1}{\bar{\alpha}}}\left(\frac{88}{60}\right)\bar{V}_{\text{MRI}} \tag{F-17}$$

where

$$\bar{V}_{\text{MRI}} = \frac{V_{\text{MRI}}}{1.52} \tag{F-18}$$

Approximate ranges in the fundamental natural frequency and damping ratio for suspension structures are given in Table F-2. The frequencies in this table are based on a limited review of typical suspension structure dynamic properties and are not intended to be applicable for every type of transmission structure. Since little data are available on damping ratios for transmission line structures, the values given in Table F-2 are conservative estimates for most structure types. The designer is encouraged to perform numerical analyses or dynamic tests in order to determine the appropriate properties. Additional information on the dynamic response of latticed towers and guyed masts can be found in ASCE (2002).

Table F-2. Approximate Dynamic Properties for Transmission Structures

Type of structure	Fundamental frequency (Hz), f_t	Damping ratio, ζ_t
Latticed tower	2.0–4.0	0.04
H-frame	1.0–2.0	0.02
Pole	0.5–1.0	0.02

Based on a past survey of transmission latticed tower frequencies, the frequency of a tower can be estimated as

$$f_t = \frac{328}{h} \tag{F-19}$$

where h is the total height of the tower (ft).

The fundamental frequency of pole-type structures can be calculated using general engineering theory. The taper dimension must be included in the estimation of the fundamental frequency.

The following equations may be used to approximate the frequency of the wires and the damping ratio of the wires if they are not known:

$$f_w = \sqrt{\frac{1}{\text{sag}}} \tag{F-20}$$

$$\zeta_w = 0.000048C_f\left(\frac{12V_o}{f_w d}\right) \tag{F-21}$$

It should be noted that Equations (F-19), (F-20), and (F-21), as well as the values in Table F-2, should be regarded as estimates only. If more accurate estimates of frequency or damping are available, such as those obtained through numerical analyses or dynamic tests, these values may be used to improve the estimation of the gust response factors.

F.4 ASSUMPTIONS AND LIMITATIONS

To derive the equations in Chapter 2, Section 2.1.5 (Eq. 2-4 and Eq. 2-5), some simplifying assumptions were made based on work carried out by Davenport (1979). These assumptions are listed below:

1. The separation coefficient, ε, is equal to 0.75 and reflects the noncoincident nature of strong wind loads on the structures and wires.
2. The statistical peak factors for background response and wind loading, g_B and g_v, are equal to 3.6. The peak factors are approximated for structures responding to buffeting wind with a broad spectrum of energy over a range of frequencies.
3. The resonant component of the dynamic response for both structure and wire systems, R_t and R_w, can be neglected for transmission structures of typical size. For typical systems, tower vibration is small due to the relatively high frequency of the structure. Wire

vibration is generally low due to the high aerodynamic damping associated with wire motion at design wind speeds. These aspects are reflected in the design equations and supported by the observation that the resonant component is often not significant in transmission lines. For tall or unique structures, designers are encouraged to consider including the resonant component of the response in the design.

F.5 EXAMPLE

The following example calculates the gust response factors for the example structure and line given in Chapter 5 of this manual. The gust response factors for the structure and the wires are calculated for (1) the background dynamic response only (simplified method in Chapter 2), and (2) the complete dynamic response (background and resonant).

The effective height of the tower is 59.3 ft, and the effective height of the wires (this example considers the conductors only) is 74 ft (conservative approach based on strong wind conditions). The total height of the tower is 89 ft, and the wind span is 1,500 ft. The diameter of the conductors is 1.165 inches, and the estimated sag of the conductor is 36 ft. A structural damping of 0.03 (3%) is assumed for the tower. The basic wind speed for the example in this appendix is $V_{100} = 96$ mph. The tower is in terrain characteristic of Exposure Category C.

Structure Gust Response Factor, G_t

1. Background dynamic response only.

The equation for the simplified gust response factor [Equation (F-8)] is

$$G_t = \left(\frac{1 + 4.6 I_z B_t}{1 + 6.1 I_z} \right)$$

From Table F-1, the wind parameters for Exposure C are

Table F-1. Wind Parameters by Exposure Category.

Exposure	α	$\overline{\alpha}$	c_{exp}	L_s (ft)	z_g (ft)
C	9.5	6.5	0.2	220	900

From Equation (F-2) $$I_z = c_{\exp}\left(\frac{33}{z_h}\right)^{\frac{1}{6}}$$

$$= 0.2\left(\frac{33}{59.3}\right)^{\frac{1}{6}}$$

$$\mathbf{= 0.18}$$

From Equation (F-3) $$g_b = g_v = 3.6$$

From Equation (F-5) $$\varepsilon = 0.75$$

From Equation (F-6) $$B_t = \sqrt{\frac{1}{1+\frac{0.56z_h}{L_s}}}$$

$$= \sqrt{\frac{1}{1+\frac{0.56(59.3)}{220}}}$$

$$\mathbf{= 0.932}$$

From Equation (F-8) $$G_t = \left(\frac{1+4.6I_zB_t}{1+6.1I_z}\right)$$

$$= \left(\frac{1+4.6(0.181)(0.932)}{1+6.1(0.181)}\right)$$

$$\mathbf{= 0.844}$$

2. Consider the complete dynamic response.
The equation for the detailed gust response factor [Equation (F-1)] is

$$G_t = \left(\frac{1+1.7I_z\varepsilon\sqrt{g_B^2B_t^2+g_{Rt}^2R_t^2}}{1+1.7I_zg_v}\right)$$

The parameters I_z, ε, g_B, g_v, and B_t are as for the background response.

From Equation (F-19), the frequency of the tower can be estimated based on a full tower height, h, of 89 ft

$$f_t = \frac{328}{h} = \frac{328}{89} = 3.69 \text{ Hz}$$

$$\zeta_t = 0.03$$

From Equation (F-4) $g_{Rt} = \sqrt{2\ln(3600 f_t)} + \dfrac{0.577}{\sqrt{2\ln(3600 f_t)}}$

$$= \sqrt{2\ln(3600 \times 3.69)} + \frac{0.577}{\sqrt{2\ln(3600 \times 3.69)}}$$

$$= 4.49$$

From Equation (F-18) $\bar{V}_{MRI} = \dfrac{V_{MRI}}{1.52} = \dfrac{96}{1.52} = 63.2 \text{ mph}$

From Equation (F-17) $V_o = 1.66\left(\dfrac{z_h}{z_g}\right)^{\frac{1}{\bar{\alpha}}}\left(\dfrac{88}{60}\right)\bar{V}_{MRI}$

$$= 1.66\left(\frac{59.3}{900}\right)^{\frac{1}{6.5}}\left(\frac{88}{60}\right)(63.2)$$

$$= 101.3 \text{ ft/s}$$

From Equation (F-7) $R_t = \sqrt{\left(\dfrac{0.0123}{\zeta_t}\right)\left(\dfrac{f_t z_h}{V_o}\right)^{\frac{-5}{3}}}$

$$= \sqrt{\left(\frac{0.0123}{0.03}\right)\left(\frac{(3.69)(59.3)}{101.3}\right)^{\frac{-5}{3}}}$$

$$= 0.337$$

Substituting solved values in Equation (F-1)

$$G_t = \left(\frac{1 + 1.7 I_z \varepsilon \sqrt{g_B^2 B_t^2 + g_{Rt}^2 R_t^2}}{1 + 1.7 I_z g_v} \right)$$

$$= \left(\frac{1 + 1.7(0.181)(0.75)\sqrt{(3.6)^2 (0.932)^2 + (4.49)^2 (0.337)^2}}{1 + 1.7(0.181)(3.6)} \right)$$

$$= \mathbf{0.877}$$

Note that, in this example, the gust response factor considering the complete dynamic response is about 4% greater than that calculated with the simplified equations in Chapter 2 (which consider the background component only).

Wire Gust Response Factor, G_w

1. Background dynamic response only

The equation for the simplified gust response factor [Equation (F-16)] is

From Equation (F-10) $\quad I_z = c_{\exp} \left(\frac{33}{z_h} \right)^{\frac{1}{6}}$

$$= 0.2 \left(\frac{33}{74} \right)^{\frac{1}{6}}$$

$$= \mathbf{0.175}$$

From Equation (F-11) $\quad g_b = g_v = 3.6$

From Equation (F-13) $\quad \varepsilon = 0.75$

From Equation (F-14) $\quad B_w = \sqrt{\frac{1}{1 + \frac{0.8S}{L_s}}}$

$$= \sqrt{\frac{1}{1+\frac{0.8(1500)}{220}}}$$

$$\mathbf{=0.394}$$

From Equation (F-16) $G_w = \left(\frac{1+4.6I_z B_w}{1+6.1I_z}\right)$

$$= \left(\frac{1+4.6(0.177)(0.394)}{1+6.1(0.177)}\right)$$

$$\mathbf{= 0.637}$$

2. Consider the complete dynamic response.
The equation for the detailed gust response factor [Equation (F-9)] is

$$G_w = \left(\frac{1+1.7I_z\varepsilon\sqrt{g_B^2B_w^2 + g_{Rw}^2R_w^2}}{1+1.7I_z g_v}\right)$$

The parameters I_z, ε, g_B, g_v, and B_w are as for the background response only.

The frequency of the conductor wire can be estimated based on a sag of the wire of 36 ft

$$f_w = \sqrt{\frac{\mathbf{1}}{\mathbf{sag}}} = \sqrt{\frac{\mathbf{1}}{\mathbf{36}}} = \mathbf{0.167}$$

From Equation (F-21), the damping of the conductor wire can be estimated based on wire and wind parameters

$$V_o = 1.66\left(\frac{z_h}{z_g}\right)^{\frac{1}{\bar{\alpha}}}\left(\frac{88}{60}\right)\bar{V}_{\mathrm{MRI}}$$

$$= 1.66\left(\frac{74}{900}\right)^{\frac{1}{6.5}}\left(\frac{88}{60}\right)(63.2)$$

$$\mathbf{= 104.8\ ft/s}$$

$$\zeta_w = 0.000048C_f\left(\frac{12V_o}{f_w d}\right)$$

$$= 0.000048(1.0)\left(\frac{12(104.8)}{(0.167)(1.165)}\right)$$

= 0.310

From Equation (F-12) $g_{Rw} = \sqrt{2\ln(3600f_w)} + \frac{0.577}{\sqrt{2\ln(3600f_w)}}$

$$= \sqrt{2\ln(3600 \times 0.167)} + \frac{0.577}{\sqrt{2\ln(3600 \times 0.167)}}$$

= 3.74

From Equation (F-15) $R_w = \sqrt{\left(\frac{0.0113}{\zeta_w}\right)\left(\frac{z_h}{S}\right)\left(\frac{f_w z_h}{V_o}\right)^{\frac{-5}{3}}}$

$$= \sqrt{\left(\frac{0.0113}{0.310}\right)\left(\frac{74}{1500}\right)\left(\frac{(0.167)(74)}{104.8}\right)^{\frac{-5}{3}}}$$

= 0.252

Substituting solved values in Equation (F-9)

$$G_w = \left(\frac{1 + 1.7I_z\varepsilon\sqrt{g_B^2B_w^2 + g_{Rw}^2R_w^2}}{1 + 1.7I_z g_v}\right)$$

$$= \left(\frac{1 + 1.7(0.175)(0.75)\sqrt{(3.6)^2(0.394)^2 + (3.74)^2(0.252)^2}}{1 + 1.7(0.175)(3.6)}\right)$$

= 0.666

Note that in this case, the gust response factor considering the complete dynamic response is about 5% greater than that calculated with the simplified equations in Chapter 2 (which consider the background component only).

APPENDIX G
SUPPLEMENTAL INFORMATION ON FORCE COEFFICIENTS

G.1 CONDUCTOR AND SHIELD WIRE FORCE COEFFICIENTS

Wind tunnel test data, such as those shown in Figure G-1, indicate that measured force coefficients for stranded wires show a wide range of variation depending on Reynolds number and the type of stranding. For this reason, there is also a wide variation in values recommended by various design codes and guides as illustrated in Figure G-2.

A force coefficient of 1.0 is recommended in Chapter 2, Section 2.1.6.2 for all conductors and shield wires. This is the same value recommended in NESC (2012). The data in Figure G-1 indicate that the force coefficient can be significantly greater than 1.0, particularly for Reynolds numbers less than 3×10^4 (small wires under nominal wind speed). For Reynolds numbers above this value, the force coefficients are reduced to a value of 1.0 or less. The expression for Reynolds number is given in Equation (2-9).

For a 0.5 inch diameter wire or larger, the Reynolds number will exceed 3×10^4 for the range of design wind speeds given in Chapter 2, Figure 2-1. For this reason, a value of 1.0 has been chosen for all conductors and shield wires. However, force coefficients larger than 1.0 are often appropriate, especially on wires having a small diameter (< 0.5 inch) and wires having accreted ice.

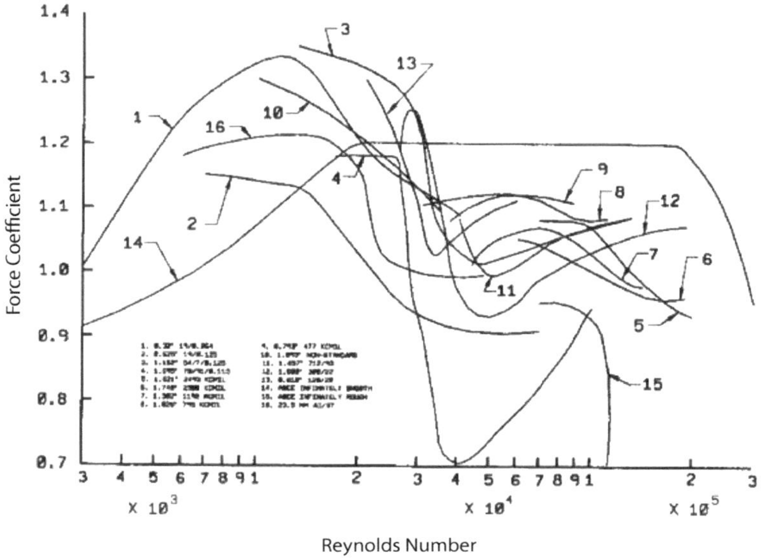

Figure G-1. Force coefficients for conductors based on wind tunnel tests.
Source: Data from ASCE (1961), Birjulin et al. (1960), Castanheta (1970), Engleman and Marihugh (1970), Richards (1965), and Watson (1955).

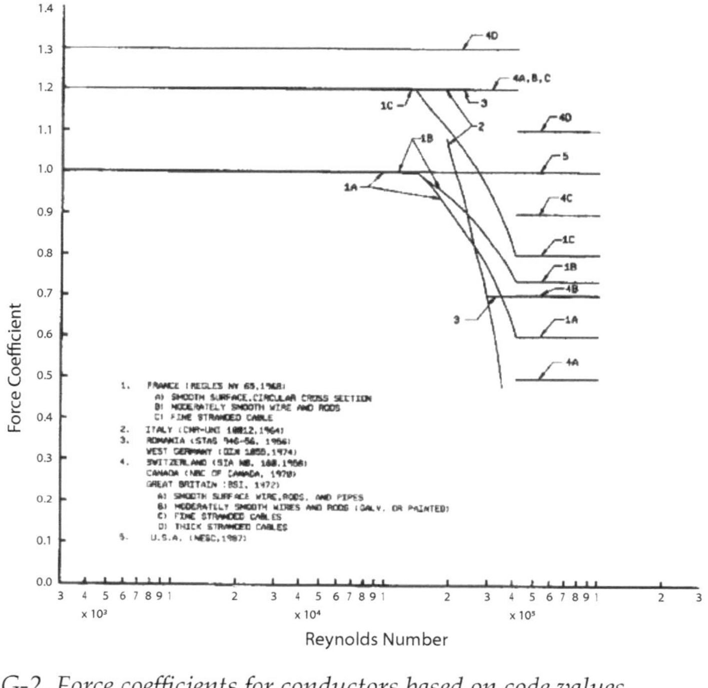

Figure G-2. Force coefficients for conductors based on code values.

G.2 MEMBER FORCE COEFFICIENTS

Table 2-5 lists recommended force coefficients for some common structural shapes used in transmission structures. Table G-1 lists force coefficients from various sources for these members and for additional shapes not listed in Table 2-5. For some shapes, values are given for variations in surface roughness, Reynolds number, corner radius ratio, yaw angle, or test conditions.

The force coefficients of asymmetrical shapes are dependent on the orientation of the wind with respect to the cross section of the member. No general equation exists for this condition; however, values have been determined through wind tunnel testing. These instances are indicated in Table G-1.

Table G-1. Member Force Coefficients.

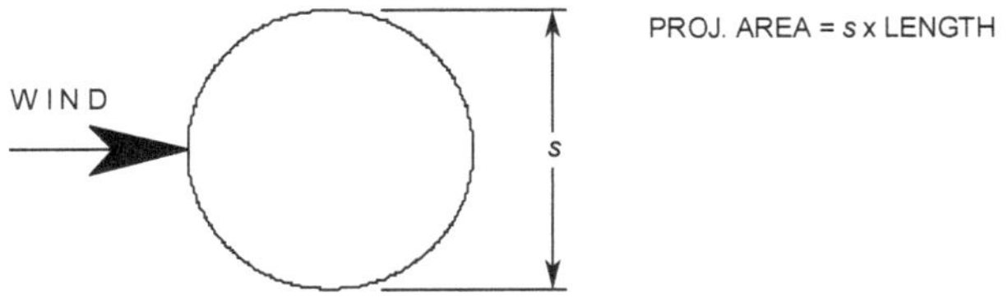

Circle

Surface	Reynolds number	Force coefficient	Reference
Any	$< 3.5 \times 10^5$	1.2	Scruton and Newberry (1963)
Any	$< 4.1 \times 1^{05}$	1.2	MacDonald (1975)
Smooth	—	0.7	ASCE (1990a)
Smooth	$< 10^5$	1.0	Sachs (1978)
Smooth	$< 3.0 \times 10^5$	1.1	AASHTO (1975)
Smooth	$> 3.5 \times 10^5$	0.7	Scruton and Newberry (1963)
Smooth	$> 4.1 \times 10^5$	0.6	MacDonald (1975)
Smooth	$3 \times 10^5 < Re < 6 \times 10^5$	$14.5 \times 10^6/Re1.3$	AASHTO (1975)
Smooth	$> 6.0 \times 10^5$	0.45	AASHTO (1975)
Rough	$> 4.1 \times 10^5$	1.2	MacDonald (1975)
Rough	—	0.9	ASCE (1990a)
Very rough	$> 3.5 \times 10^5$	1.0	Scruton and Newberry (1963)
Very rough	—	1.2	ASCE (1990a)

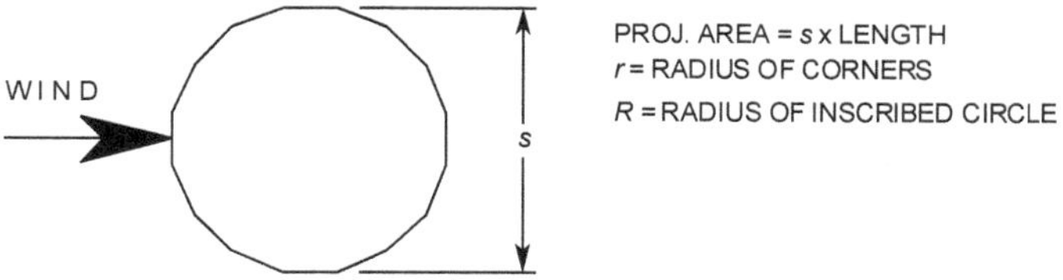

16-sided polygon

Corner radius (r/R)	Reynolds number	Force coefficient	Reference
< 0.26	$> 6.0 \times 10^5$	$0.83–1.08(r/R)$	James (1976)
> 0.26	$> 6.0 \times 10^5$	0.55	James (1976)

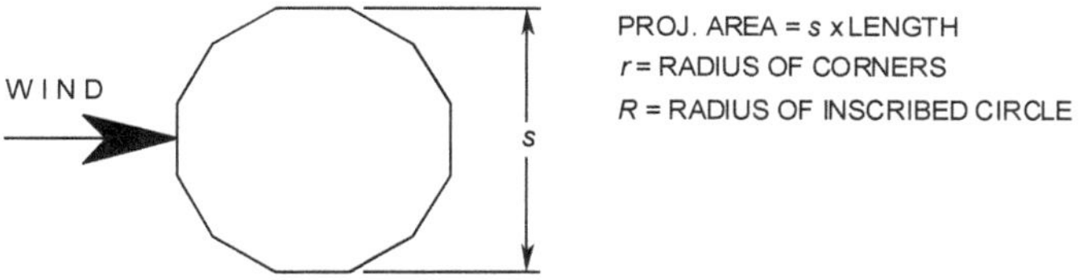

12-sided polygon

Corner radius (r/R)	Reynolds number	Force coefficient	Reference
0	$< 3.5 \times 10^5$	1.3	Scruton and Newberry (1963)
0	$< 8.2 \times 10^5$	1.3	MacDonald (1975)
0	$> 3.5 \times 10^5$	1.0	Scruton and Newberry (1963)
0	$> 8.2 \times 10^5$	1.1	MacDonald (1975)
$0.09 < r/R < 0.34$	$> 10^6$	$0.936–1.087(r/R)$	James (1976)
> 0.125	$< 3.0 \times 10^5$	1.2	AASHTO (1975)
> 0.125	$3.0 \times 10^5 < Re < 6.0 \times 10^5$	$2{,}322/Re^{0.6}$	AASHTO (1975)
> 0.125	$> 6.0 \times 10^5$	0.79	AASHTO (1975)
> 0.34	$> 10^6$	0.57	James (1976)

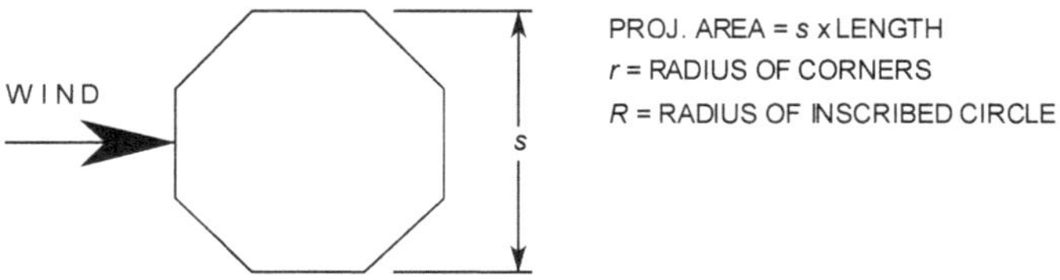

8-sided polygon

Corner radius (r/R)	Reynolds number	Force coefficient	Reference
0	—	1.2	AASHTO (1975)
0	—	1.4	ASCE (1990a), MacDonald (1975)
$0.09 < r/R < 0.59$	$> 10^6$	$1.422–1.368(r/R)$	James (1976)
> 0.59	$> 10^6$	$0.744–0.194(r/R)$	James (1976)

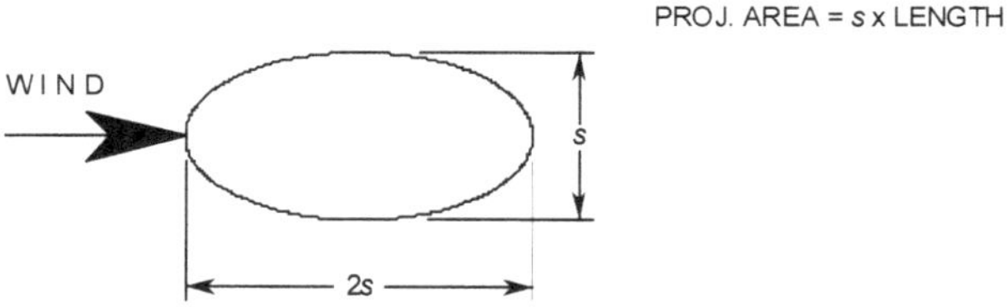

Ellipse, wind on narrow side

Sides	Reynolds number	Force coefficient	Reference
Smooth	$< 6.9 \times 10^5$	0.7	MacDonald (1975)
Smooth	$> 6.9 \times 10^5$	0.2	MacDonald (1975)
Multi-sided	—	$(C/3)(4 - D/d)$	AASHTO (1975)
where	D = Major diameter d = Minor diameter $D/d = 2.0$ C = Force coefficient of cylindrical shape with diameter equal to D		

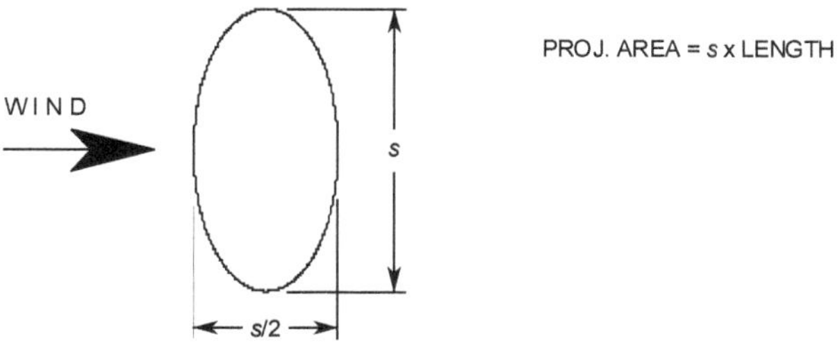

Ellipse, wind on broad side

Sides	Reynolds number	Force coefficient	Reference
Smooth	$< 5.5 \times 10^5$	1.7	MacDonald (1975)
Smooth	$> 5.5 \times 10^5$	1.5	MacDonald (1975)
Multisided	—	$1.7(D/d - 1) + C(2 - D/d)$	AASHTO (1975)

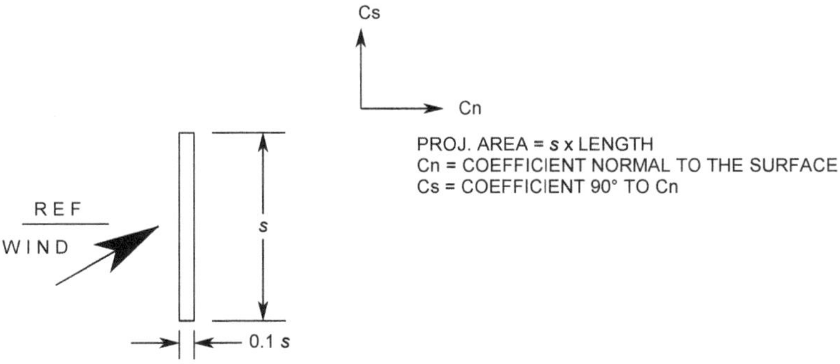

Flat plate

Angle	C_n	C_s	Reference
0°	2.0	0.0	Scruton and Newberry (1963), Sachs (1978)
45°	1.8	0.1	Sachs (1978)
90°	0.0	0.1	Sachs (1978)

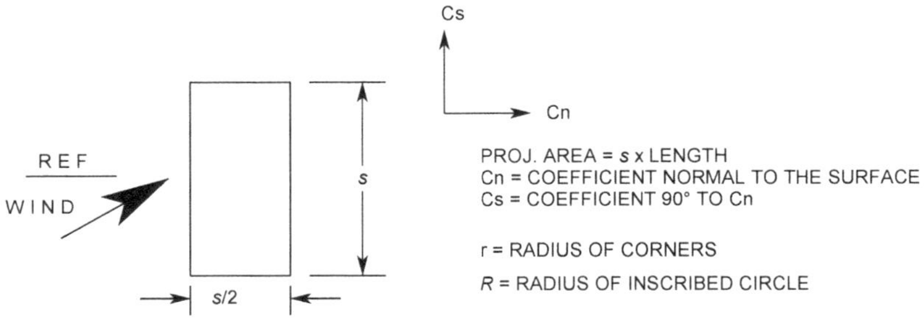

Rectangle

Corner radius (r/R)	Angle	C_n	C_s	Reference
0	0°	2.2	0.0	Scruton and Newberry (1963)
0	0°	2.1	0.0	Sachs (1978)
0	45°	1.4	0.7	Sachs (1978)
0	90°	0.0	0.75	Sachs (1978)
0.08	0°	1.9	0.0	MacDonald (1975)
0.25	0°	1.6	0.0	Scruton and Newberry (1963)

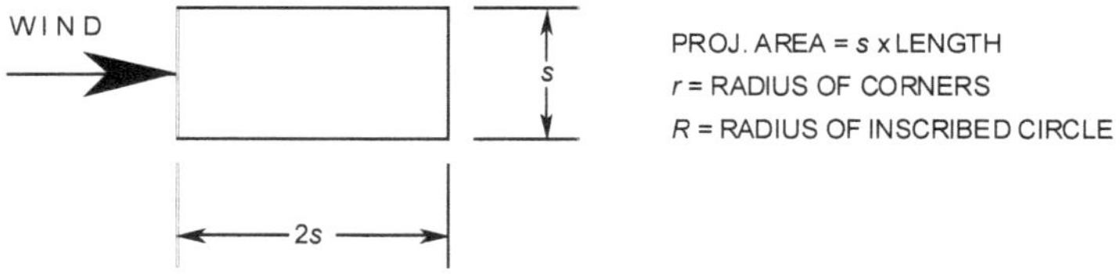

Rectangle

Corner radius (r/R)	Reynolds number	Force coefficient	Reference
0.0	—	1.4	Scruton and Newberry (1963)
0.167	—	0.7	MacDonald (1975)
0.5	—	0.4	Sachs (1978)

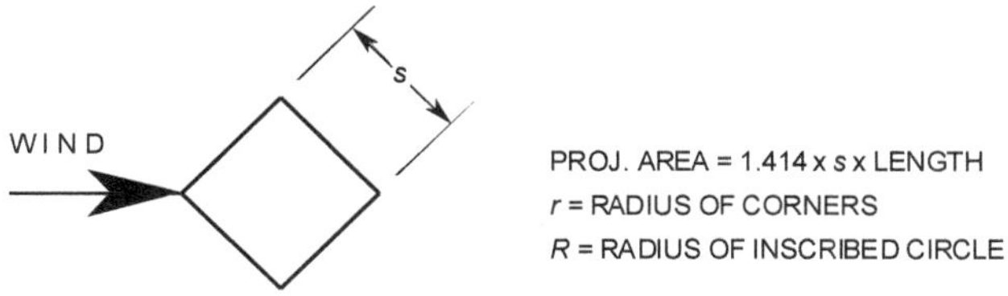

Square, wind at apex (cornering)

Corner radius (r/R)	Reynolds number	Force coefficient	Reference
0.0	—	1.5	ASCE (1990b), Scruton and Newberry (1963)
0.33	$< 6.86 \times 10^5$	1.5	MacDonald (1975)
0.33	$> 6.86 \times 10^5$	0.6	MacDonald (1975)

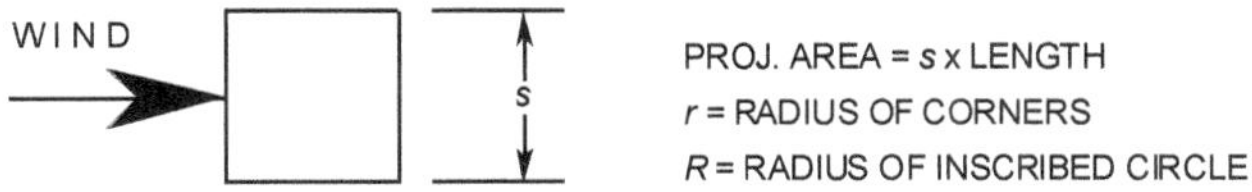

Square, wind at side

Corner radius (r/R)	Reynolds number	Force coefficient	Reference
0.0	—	2.0	ASCE (1990b), Scruton and Newberry (1963)
0.167	$< 6.86 \times 10^5$	1.3	MacDonald (1975)
0.167	$> 6.86 \times 10^5$	0.6	MacDonald (1975)
0.33	$< 2.7 \times 10^5$	1.0	MacDonald (1975)
0.33	$> 2.7 \times 10^5$	0.5	MacDonald (1975)

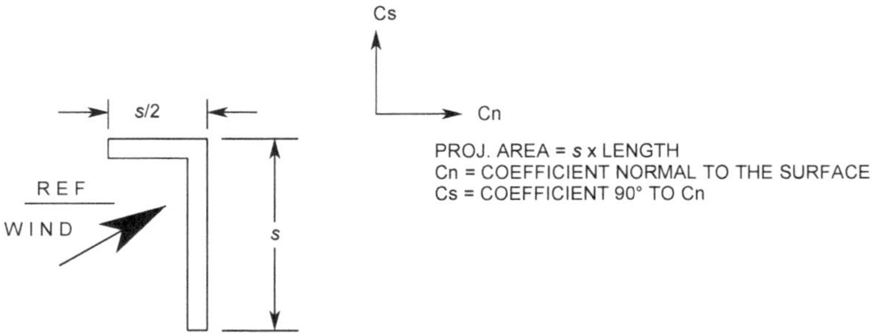

Unequal leg angle

Angle	C_n	C_s	Reference
0°	1.9	0.95	Sachs (1978)
45°	1.8	0.8	Sachs (1978)
90°	2.0	1.7	Sachs (1978)
135°	–1.8	–0.1	Sachs (1978)
180°	–2.0	0.1	Sachs (1978)

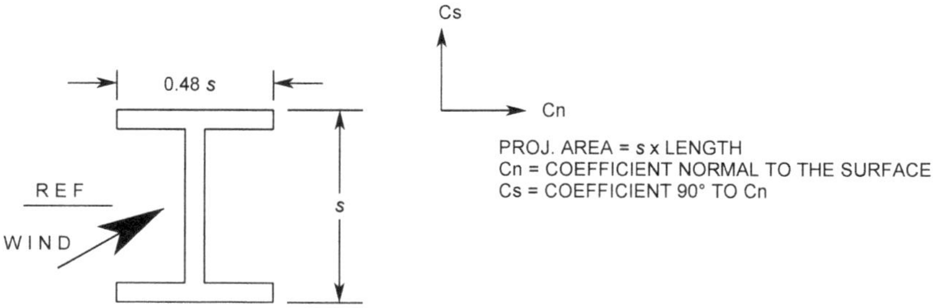

I-beam

Angle	C_n	C_s	Reference
0°	2.05	0.0	Sachs (1978)
45°	1.95	0.6	Sachs (1978)
90°	0.5	0.9	Sachs (1978)

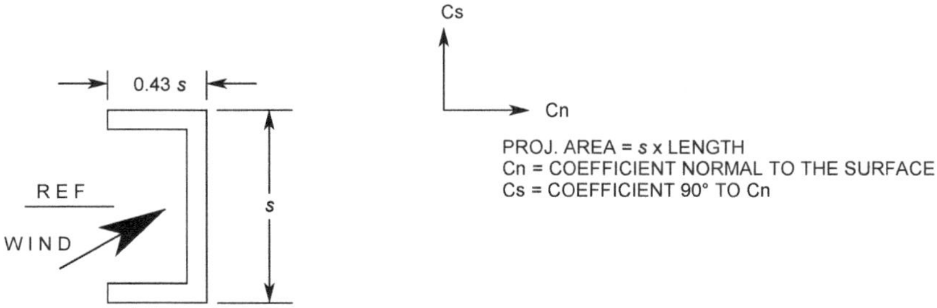

Channel

Angle	C_n	C_s	Reference
0°	2.05	0.0	Sachs (1978)
45°	1.85	0.6	Sachs (1978)
90°	0.0	0.6	Sachs (1978)
135°	–1.6	0.4	Sachs (1978)
180°	–1.8	0.0	Sachs (1978)

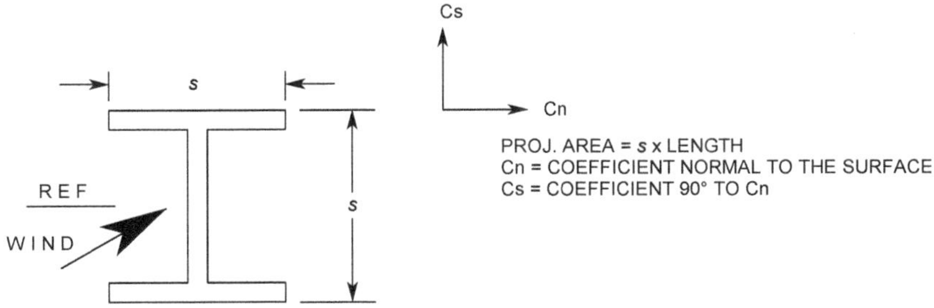

Wide flange

Angle	C_n	C_s	Reference
0°	1.6	0.0	Sachs (1978)
45°	1.5	1.5	Sachs (1978)
90°	0.0	1.9	Sachs (1978)

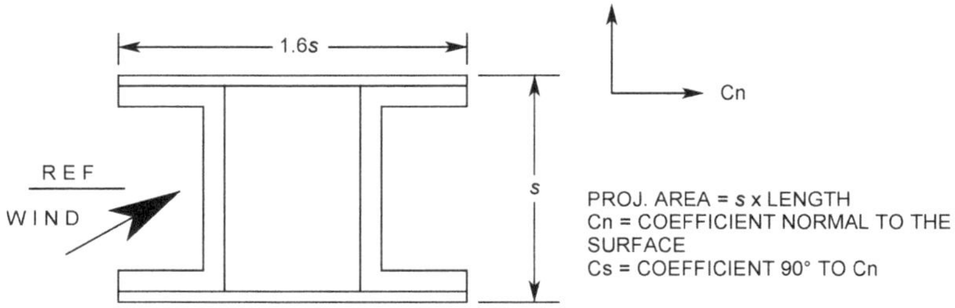

Built-up section

Angle	C_n	C_s	Reference
0°	1.4	0.0	Sachs (1978)
45°	1.2	1.6	Sachs (1978)
90°	0.0	2.2	Sachs (1978)

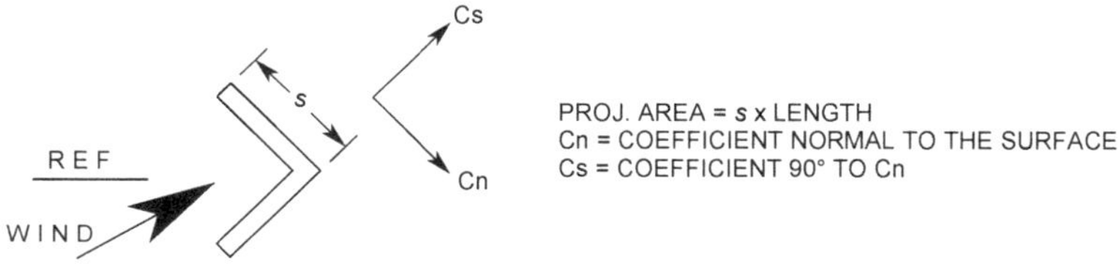

Equal leg angle

Angle	C_n	C_s	Reference
0°	1.8	1.8	Sachs (1978)
45°	2.1	1.8	Sachs (1978)
90°	–1.9	–1.0	Sachs (1978)
135°	–2.0	0.3	Sachs (1978)
180°	–1.4	–1.4	Sachs (1978)

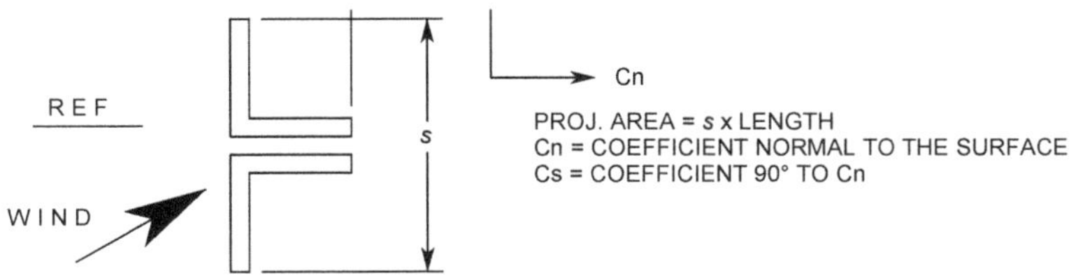

Double angle

Angle	C_n	C_s	Reference
0°	1.6	0.0	Sachs (1978)
45°	1.5	–0.1	Sachs (1978)
90°	–0.95	0.7	Sachs (1978)
135°	–0.5	1.05	Sachs (1978)
180°	–1.5	0.0	Sachs (1978)

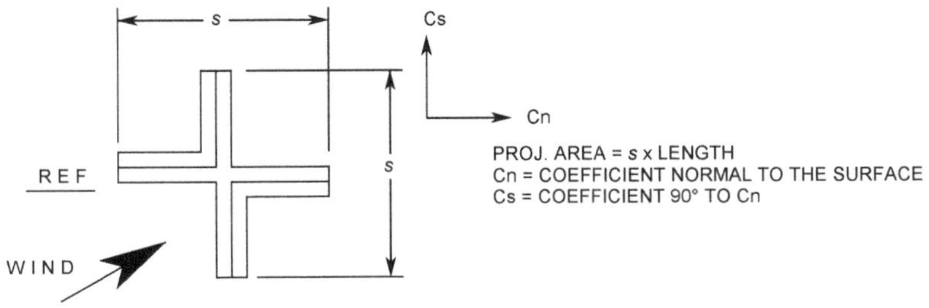

Built-up angles

Angle	C_n	C_s	Reference
0°	1.75	0.1	Sachs (1978)
45°	0.85	0.85	Sachs (1978)
90°	–0.1	1.75	Sachs (1978)
135°	–0.75	0.75	Sachs (1978)
180°	–1.75	–0.1	Sachs (1978)

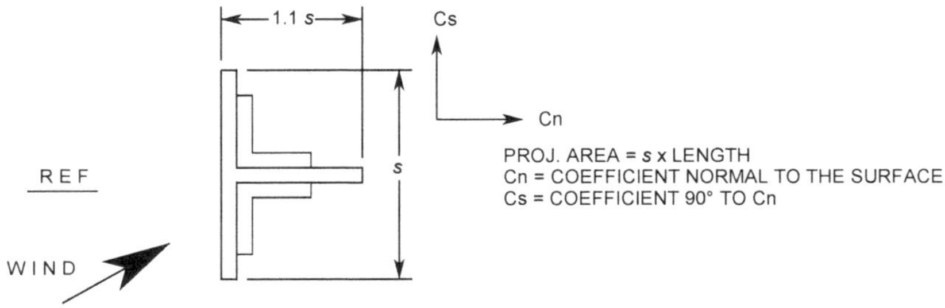

T-section

Angle	C_n	C_s	Reference
0°	2.0	0.0	Sachs (1978)
45°	1.2	0.9	Sachs (1978)
90°	–1.6	2.15	Sachs (1978)
135°	–1.1	2.4	Sachs (1978)
180°	–1.7	2.1	Sachs (1978)

G.3 ASPECT RATIO

The force coefficients given in Chapter 2, Section 2.1.6.2 and in Section G.2 are for infinitely long members and are applicable to members with aspect ratios greater than 40. Adjustment factors for members with aspect ratios less than 40 may be applied as follows (MacDonald 1975):

$$C_f' = (c)(C_f) \tag{G-1}$$

where

c = Correction factor for aspect ratio (Table G-2),
C_f = Force coefficient from Section 2.1.6.2 or G.2, and
C_f' = Force coefficient corrected for aspect ratio.

Table G-2. Aspect Ratio Correction Factors

Aspect ratio	Correction factor (c)
0–4	0.6
4–8	0.7
8–40	0.8
> 40	1

Note: Aspect ratio = (L_m/d_s) except for members attached to the ground where aspect ratio = ($2L_m/d_s$), in which L_m = member length and d_s = member diameter or width.

G.4 LATTICED TRUSS STRUCTURE FORCE COEFFICIENTS

The force coefficients calculated using Table 2-4 and Equation (2-13) in Chapter 2 represent the recommended values for square-section and triangular-section latticed structures having flat-sided and rounded members. The recommended force coefficients, which are taken directly from ASCE 7-16 (2017), account for the wind forces acting on the windward and leeward faces of the latticed tower. Therefore, they are influenced by the solidity ratio, which is defined in Equation (2-10). As the solidity ratio increases, the force coefficient is reduced due to the shielding effect of the members in the windward face(s) of the tower.

Figures G-3 through G-6 provide information from various other codes, standards, and tests for force coefficients for latticed towers with wind normal to a face. These figures are for towers having either square or triangular cross-sections and comprised of flat-sided or rounded members.

Figures G-7 through G-10 provide information from various codes and standards for force coefficients for latticed towers with yawed wind. These figures are for latticed tower structures having either square or triangular cross-sections and comprised of flat-sided or rounded members. Whitbread (1979) has published other data relating to wind forces on latticed towers having a wide variety of shapes, solidity ratios, and wind directions. The variation of the force coefficient with yaw angle was examined for square sections by Bayar (1986) and for a typical cross-arm section by Mara et al. (2010).

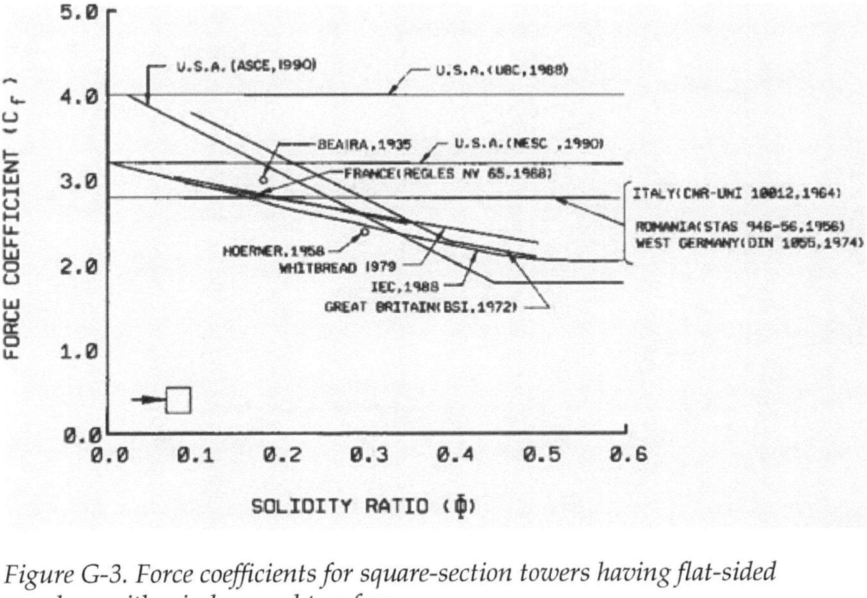

Figure G-3. Force coefficients for square-section towers having flat-sided members with wind normal to a face.

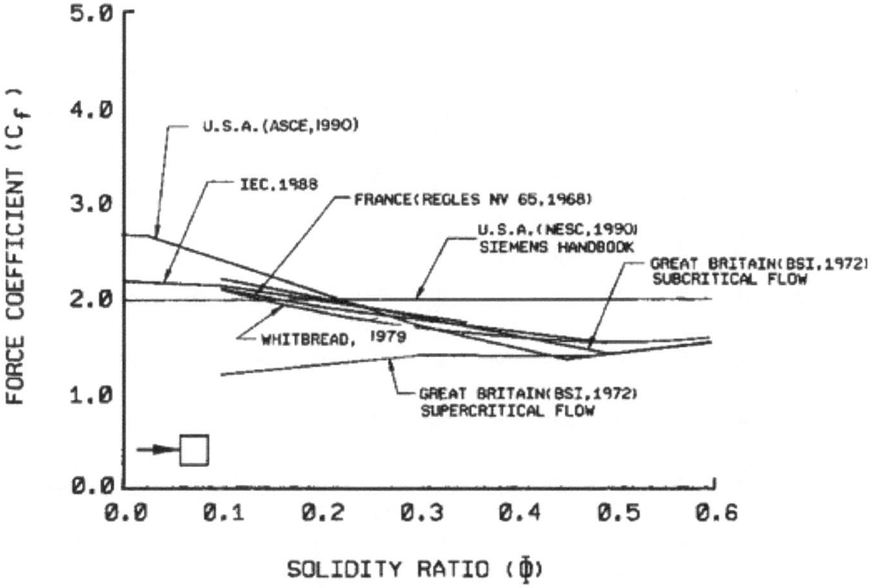

Figure G-4. Force coefficients for square-section towers having rounded members with wind normal to a face.

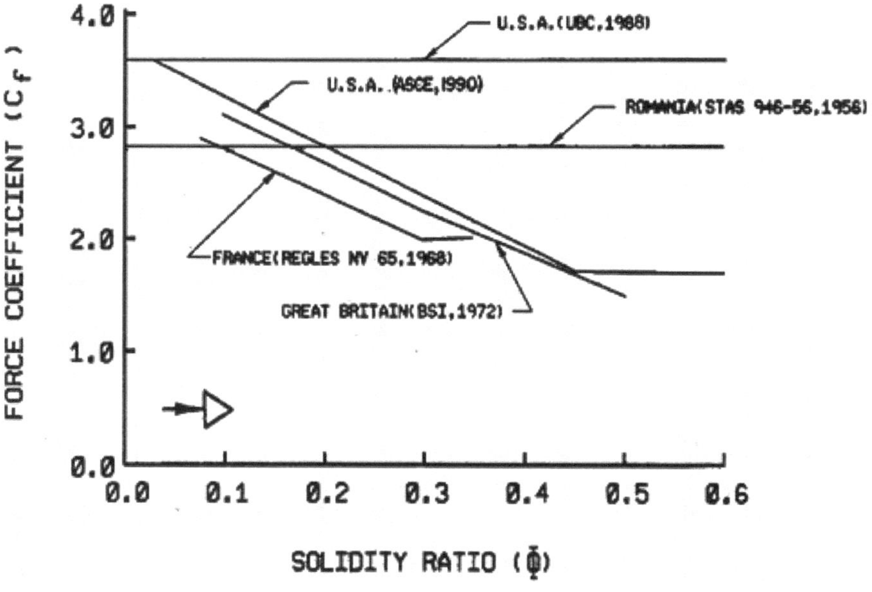

Figure G-5. Force coefficients for equilateral triangular-section towers having flat-sided members with wind normal to a face.

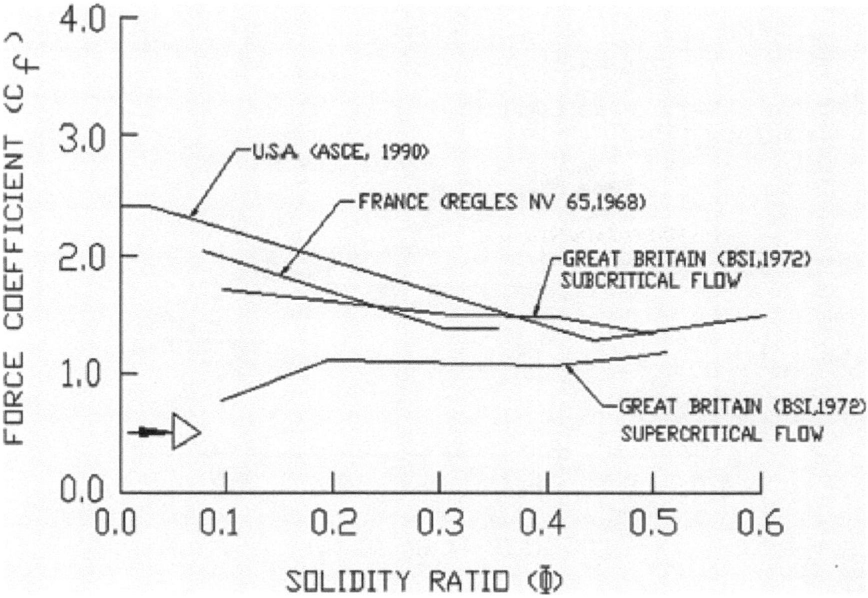

Figure G-6. Force coefficients for equilateral triangular-section towers having rounded members with wind normal to a face.

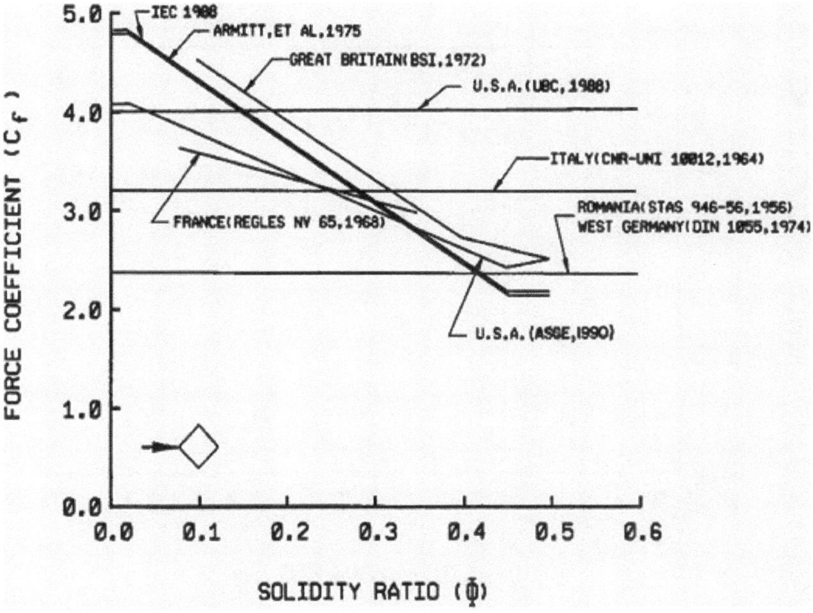

Figure G-7. Force coefficients for square-section towers having flat-sided members with diagonal wind.

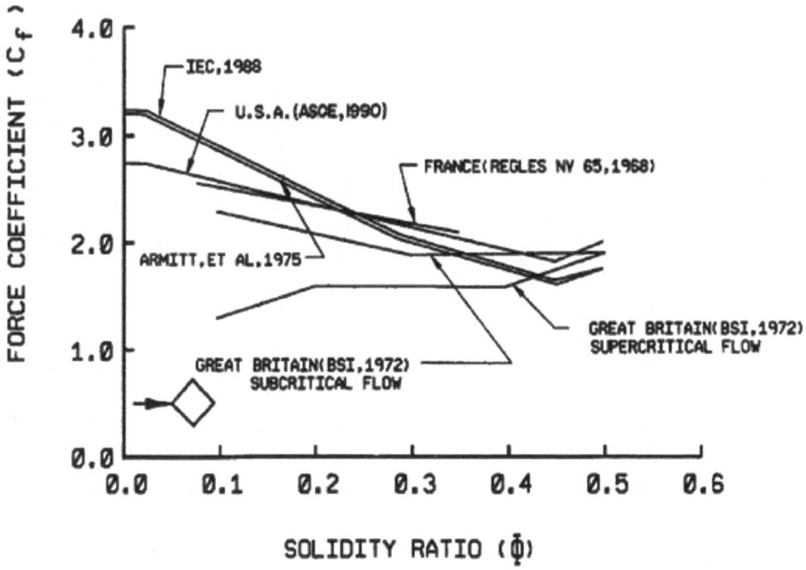

Figure G-8. Force coefficients for square-section towers having rounded members with diagonal wind.

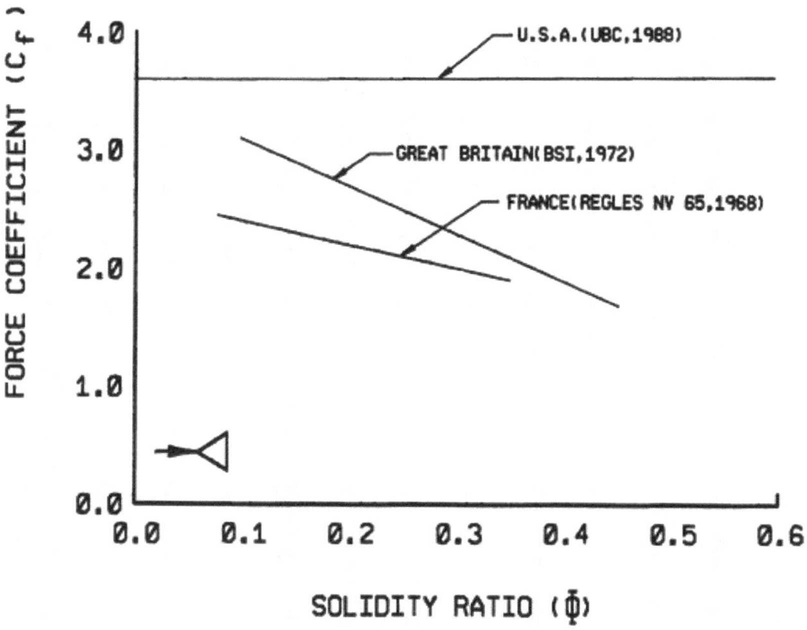

Figure G-9. Force coefficients for equilateral triangular-section towers having flat-sided members with cornering wind.

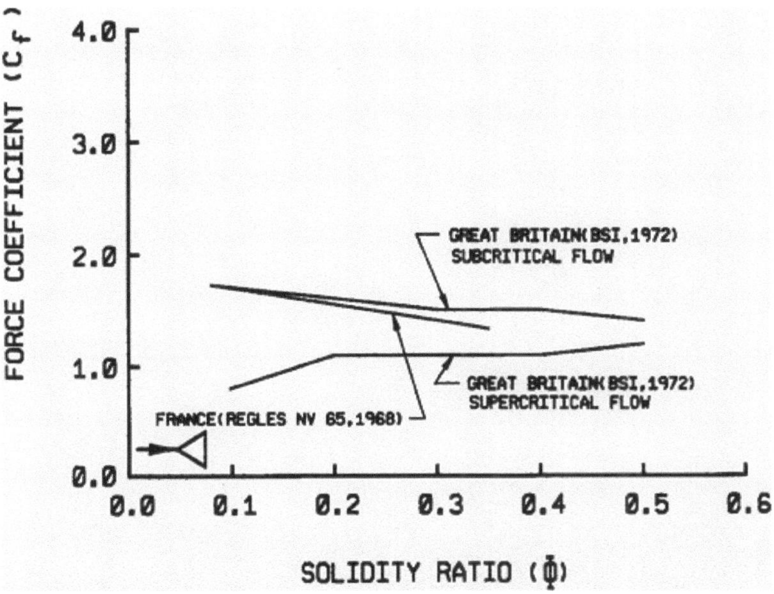

Figure G-10. Force coefficients for equilateral triangular-section towers having rounded members with cornering wind.

G.5 FORCE COEFFICIENTS OF ICED COMPONENTS

When calculating forces due to wind on ice-covered wires, the force coefficient is dependent on the shape of the ice buildup (McComber et al. 1982). However, typical force coefficients of ice-covered wires are not known. Some organizations recommend using force coefficients other than 1.0 for wires covered with ice (IEC 2003, ISO Standard 12494).

APPENDIX H
SUPPLEMENTAL INFORMATION ON ICE LOADING

H.1 THEORY AND CONDITIONS OF ICE FORMATION

In a general sense, the meteorological parameters that influence the type and amount of ice that forms under different conditions are well known. Liquid water content of supercooled clouds and precipitation intensity for freezing precipitation icing and sticky snow determine the amount of water available for ice formation. The ice properties are determined by the air temperature, wind speed, drop size, and supercooled liquid water content of clouds, fog, or precipitation intensity and type. The icing phenomenon is best classified by the causal meteorological conditions. In the following paragraphs, the various icing mechanisms are described as it is important for the engineer to understand the conditions which may result in severe loads on transmission lines.

H.1.1 Precipitation Icing

Freezing rain (or drizzle) is a common icing mechanism. Freezing rain occurs when warm, moist air is forced over a layer of subfreezing air at the Earth's surface. The precipitation usually begins as snow, which melts as it descends through the layer of warm air aloft. The drops cool as they fall through the cold surface air layer and freeze on contact with structures, or the ground, to form glaze ice. Upper air data indicate that the cold surface air layer is typically between 1,000 ft (300 m) and 3,900 ft (1,200 m) thick (Young 1978), and averages approximately 1,600 ft (500 m) (Bocchieri 1980). The warm air layer aloft averages 5,000 ft (1,500 m) thick in freezing rain, but in freezing drizzle the entire temperature profile may be below 32 °F (0 °C) (Bocchieri 1980). Precipitation associated with slowly moving frontal systems can alternate between snow and freezing rain to form a composite slow-glaze accretion on structures. The density of glaze is usually assumed to be 56 to 57 pcf (900 to 917 kg/m^3).

In freezing rain, the water impingement rate is often greater than the freezing rate. The excess water starts to drip off and may freeze as icicles, resulting in a variety of accretion shapes that range from a smooth, cylindrical sheath through a crescent on the windward side with icicles hanging on the bottom to large, irregular protuberances. The shape of a glaze accretion depends on the varying meteorological factors and the cross-sectional shape of the structural member or component, its spatial orientation, and flexibility.

H.1.2 In-Cloud Icing

This icing condition occurs when supercooled cloud or fog water droplets, 100 μm or less in diameter, collide with a structure. This occurs in mountainous areas where adiabatic cooling causes saturation of the atmosphere to occur at temperatures below freezing, in free air in supercooled clouds, and in supercooled fogs that exist in a stable air mass caused by a strong temperature inversion. Significant accumulations of ice can result. Large concentrations of supercooled droplets are not common at air temperatures below about 0 °F (−18 °C).

In-cloud icing forms rime or glaze ice with a density between about 10 and 56 pcf (150 and 900 kg/m^3), depending on the amount of entrapped air. If the heat of fusion that is released by the freezing droplets is removed by convective and evaporative cooling faster than it is released, the droplets freeze on impact. The degree to which the droplets spread as they collide and freeze governs how much air is incorporated in the accretion, and thus its density. If the cooling rate is relatively low, not all the colliding droplets freeze. The resulting ice accretion will be clear or opaque, possibly with attached icicles.

The collision efficiency of a structure is defined as the fraction of cloud droplets in the volume swept out by the structure that actually collide with it. The basic theory of the collision efficiency of smooth, circular cylinders perpendicular to the flow of droplets carried by a constant wind was developed by Langmuir and Blodgett (1946). Collision efficiency increases with wind speed and droplet diameter and decreases as the diameter of the cylinder increases. For a given wind speed and droplet size, the theory defines a critical cylinder diameter beyond which accretion will not occur. This concept of a critical diameter has been confirmed by observation. Formulas for calculating collision efficiencies based on an updated numerical analysis are provided in Finstad and Lozowski (1988).

The amount of ice accreted during in-cloud icing depends on the duration of the icing condition and the wind speed, as well as on the liquid water content and the size of the droplets in the supercooled clouds or fog. If, as often occurs, wind speed increases and air temperature decreases with height aboveground, larger amounts of ice will accrete on higher structures. The accretion shape depends on the flexibility of the structural member or component. If it is free to rotate, such as a long guy or a long span of a single conductor or wire, the ice accretes with a roughly circular cross section. On more rigid structural members and components, the ice forms in pennant shapes extending into the wind.

H.1.3 Snow

Sticky snow that falls on a round cross-sectional structural member or component (such as a wire, cable, conductor, or guy) may deform and/or slide around it. Due to the shear and tensile strength of the snow resulting from capillary forces, interparticle freezing (Colbeck and Ackley 1982), and/or sintering (Kuroiwa 1962), the accreting snow may not fall off the structural member during this process. Ultimately, the snow forms a cylindrical sleeve, even around bundled conductors and wires. The formation of the snow sleeve is enhanced by torsional rotation of flexible structural members or components because of the eccentric weight of the snow. The density of accreted snow ranges from below 5 to 50 pcf (80 to 800 kg/m^3) and may be much higher than the density of the same snowfall on the ground.

Damaging snow accretions have been observed at surface air temperatures ranging from the low 20 °F up to about 36 °F (−5 °C to 2 °C). Snow with high moisture content appears to stick more readily than drier snow. Snow falling at a surface air temperature above 32 °F (0 °C) may accrete even at wind speeds above 25 mph (10 m/s), producing dense [37 to 50 pcf (600 to 800 kg/m^3)] accretions. Snow with lower moisture content is not as sticky, blowing off the structure in high winds. These accreted snow densities are typically between 2.5 and 16 pcf (40 and 250 kg/m^3) (Kuroiwa 1965). Dry snow can also accrete on structures (Gland and Admirat 1986). The cohesive strength of the dry snow is initially supplied by the interlocking of the flakes, and ultimately by sintering, as molecular diffusion increases the bond area between adjacent snowflakes. These dry snow accretions appear to form only in very low winds and have densities estimated at between 5 and 10 pcf (80 and 150 kg/m^3) (Sakamoto et al. 1990, Peabody 1993).

H.1.4 Hoarfrost

Hoarfrost is an accumulation of ice crystals formed by direct deposition of water vapor from the air onto a structure. Because it forms when air with a dew point below freezing is brought to saturation by cooling, hoarfrost is often found early in the morning after a clear, cold night. It is feathery in appearance and typically accretes up to about an inch (25 mm) in thickness with very little weight. Hoarfrost does not constitute a significant loading problem.

H.2 LOADING IMBALANCES

Unbalanced loads from in-cloud icing may be significant (White 1999). Because the rime density and thickness increase with wind speed, significant differences in ice loading can occur from one span to the next where the transmission line crosses a ridge, hill, or escarpment. This can result in a severe loading imbalance on the line, particularly if adjacent span lengths are significantly different. When a transmission line is to be located in a region where in-cloud icing occurs, the engineer would benefit from consulting a meteorologist to determine the severity and extent of the ice loads. With this information, the engineer can either relocate the line to reduce the exposure or identify line sections with the greatest risk for in-cloud icing and adjust designs accordingly.

Snow accretions may shed from wires in the process of formation, before forming a cylindrical sleeve around the wire. Low-density snow accretions formed in light winds may shed when the wind speed increases. When snow sheds from some of the spans, the still-loaded spans will pull slack from the unloaded spans. This can cause significant increases in the sag of the wire in the loaded spans. Such events can create clearance violations, especially if there is deep snow on the ground, even though the associated unbalanced loads are typically small.

Variations in ice loading during precipitation icing are typically gradual along the length of a transmission line. Therefore, unequal icing of adjacent spans is not significant.

Unbalanced longitudinal loadings associated with ice dropping or unequal ice formation on adjacent spans depend on the relationships between available slack, insulator lengths, and other factors. Suggestions for the determination of unbalanced ice loads can be found in IEC 60826 (IEC 2003b) and the various national options of EN 50341 (CENELEC 2012).

H.3 ICE ACCRETION DATA AND MODELING

There are very little data in North America on equivalent uniform ice thicknesses from natural ice accretions on overhead lines. Therefore, ice loading studies often rely on mathematical models based on the physics of the various types of icing and on meteorological data (i.e., precipitation amount and type, temperature, wind speed) that are required as input to these models. Results from an ice accretion analysis typically give calculated ice thicknesses for past storms in which freezing precipitation has occurred. An extreme value analysis can then be applied to determine t_{MRI}. Wind speeds during and after periods of freezing precipitation can also be extracted from the meteorological data and analyzed to determine the wind speed to apply concurrently with t_{MRI}.

There are a number of ice accretion models available that use weather data to determine accreted ice loads, including the conservative Simple model (Jones 1998), similar to the Goodwin model (Goodwin et al. 1983), the US Army Cold Regions Research and Engineering Laboratory (CRREL) model (Jones 1996), the Makkonen model (Makkonen 1996), the Meteorological Research Institute (MRI) model (MRI 1977), and the Chaîné model (Chaîné and Castonguay 1974). The following comments provide information on the above mentioned models:

- The Simple model determines the ice thickness, t, from the amount of freezing rain and the wind speed. t does not depend on the air temperature because it is assumed that all the available precipitation freezes, and t also does not depend on the wire diameter.
- The CRREL model is less conservative than the Simple model, using a heat-balance calculation to determine how much of the impinging precipitation freezes directly to the wire and how much of the runoff water freezes as icicles. It calculates smaller ice loads than the Simple model when the air temperature is near freezing and wind speeds are relatively low; however, water that does not freeze immediately may freeze as icicles as it drips off the wire. The CRREL model requires the user to specify the diameter of the wire on which the accretion of ice is to be modeled. However, this model, like the Meteorological Research Institute and Makkonen models, shows very little dependence of ice thickness on wire diameter.
- The Meteorological Research Institute model tends to determine smaller ice loads than the CRREL model because water that does not freeze immediately is ignored, rather than being allowed to freeze to form icicles. However, in using that model or the Goodwin model, the user is required to specify the fall speed of the rain drops and the model results depend significantly on the speed that is chosen. The Meteorological Research Institute model also determines accreted snow loads and in-cloud icing loads; however, many of the significant parameters, including droplet size and liquid water content of the supercooled clouds, rime accretion density, and snow sticking fraction and snow accretion density, must be chosen by the user.
- The Makkonen model for ice accretion in freezing rain tends to be almost as conservative as the Simple model, primarily because it

assumes that a significant portion of the water that does not freeze immediately is incorporated in the accretion. Thus, there is relatively little water available to freeze as icicles.

- The Chaîné model is based on wind tunnel tests that were done by Stallabrass and Hearty (1967) to investigate sea-spray icing. A number of assumptions and extrapolations are made to mold these data into a formulation for freezing rain, and the results indicate a significant variation of uniform radial ice thickness with wire diameter.

There have been some attempts at model validation. Felin (1988) compared measured maximum ice thicknesses on cylinders of Hydro Quebec's Passive Ice Meters (PIM) with Meteorological Research Institute model results, assuming a drop fall speed of 9 mph (4.1 m/s). Yip and Mitten (1991) compared 61 PIM measurements with Chaîné, Makkonen, Meteorological Research Institute, and Goodwin model results using data at nearby weather stations. Yip (1995) used annual maximum ice thickness data from 235 PIM sites from 1974 to 1990 and compared the factored ice thicknesses to annual maxima from the Chaîné model. Jones (1998) compared the measured ice load on a horizontal cylinder in a single freezing rainstorm with Chaîné, Meteorological Research Institute, Makkonen, Simple, and CRREL model ice loads using co-located weather data. Newfoundland and Labrador Hydro et al. (CEA 1998) reported on the results of a 4-year Canadian Electrical Association (CEA) study comparing ice loads on three test spans with ice loads determined from the Chaîné, Makkonen, and Meteorological Research Institute models using weather data measured at the test spans in 22 storm events. In all these comparisons, the ice accretion models as well as the user interface between the weather data and the model and the assumptions made in determining the equivalent uniform radial ice thickness from the ice measurements were tested.

An alternative approach to using meteorological data and ice accretion models is to establish ice and wind measurement stations at several locations in the service area of the utility. The uniform radial thickness, t, can be determined from the typical cross-sectional area, A_i, of the ice accretion on a wire of diameter d such that

$$t = -\frac{d}{2} + \left(\frac{d^2}{4} + \frac{A_i}{\pi}\right)^{0.5} \tag{H-1}$$

or from the mass, m_i, of a typical ice sample of length L such that

$$t = -\frac{d}{2} + \left(\frac{d^2}{4} + \frac{m_i}{\pi \rho_i L}\right)^{0.5} \tag{H-2}$$

where ρ_i is the density of the ice.

In determining ice thicknesses for transmission lines from such data, the height above ground and orientation of the ice samples to the wind must be considered. With a sufficiently long period of record and a representative geographic distribution of these stations, extreme ice loads and concurrent wind speeds can be determined.

H.4 EXTREME ICE THICKNESS FROM FREEZING RAIN AND CONCURRENT WIND SPEEDS

The map of 100-year MRI ice thicknesses from freezing precipitation with concurrent wind speeds (Figures 2-19 through 2-23 in Chapter 2) was developed from the same data as the 50-year MRI map in ASCE 7-10 (2010b) and the 500-year MRI map in ASCE 7-16 (2017).

H.4.1 Continental United States and Alaska

Historical weather data from 500 National Weather Service (NWS), military, Federal Aviation Administration (FAA), and Environment Canada weather stations were used with the CRREL and Simple models to estimate glaze ice loads in past freezing rainstorms on wires 33 ft (10 m) above ground, at an orientation perpendicular to the wind. The station locations are shown in Figure H-1 for the continental United States and in Figure 2-23 for Alaska. The period of record of the meteorological data at any station is typically 20 to 50 years.

Accreted ice was assumed to remain on the wire until after freezing rain ceases and the air temperature increases to at least 33 °F (0.6 °C). The maximum ice thickness and the maximum wind-on-ice load were determined for each storm. Severe storms, such as those with significant ice or wind-on-ice loads at one or more weather stations, were researched in *Storm Data* (NOAA, 1959–present; a monthly publication that describes damage from storms of all sorts throughout the United States), newspapers, and utility reports to obtain corroborating qualitative information on the extent of damage from the storm. Very little corroborating information was obtained about damaging freezing rainstorms in Alaska, perhaps because of the low population density and relatively sparse newspaper coverage in the state. Extreme ice thicknesses were then determined using the peaks-over-threshold method and the generalized Pareto distribution (Abild et al. 1992, Hoskings and Wallis 1987, Wang 1991). To reduce sampling error, weather stations were grouped into superstations (Peterka 1992) based on the incidence of severe storms, the frequency of freezing rainstorms, latitude, proximity to large bodies of water, elevation, and terrain. A few stations that were judged to have unique freezing rain climatology were not incorporated in superstations. Concurrent wind-on-ice speeds were back-calculated from the extreme wind-on-ice load and the extreme ice thickness. In calculating wind-on-ice loads, engineers should keep in mind that the actual projected area of a glaze ice accretion may be significantly larger than that obtained by assuming a uniform ice thickness. Thus, the assumption of a force coefficient of 1.0 for an ice-covered wire will not be conservative.

Figures 2-19 through 2-23 represent the most consistent and best available nationwide maps for design ice loads. The icing model used to produce the map has not, however, been verified with a large set of co-located measurements of meteorological data and ice thicknesses. Furthermore, the weather stations used to develop this map are almost all located at airports. Structures in more exposed locations at higher elevations, or in valleys or gorges (for example, Signal and Lookout Mountains in Tennessee, the Pontotoc Ridge and the edge of the Yazoo Basin in Mississippi, the Shenandoah Valley and Poor Mountain in Virginia, Mount Washington in New Hampshire, and Buffalo Ridge in Minnesota and South Dakota) may be subject to larger ice thicknesses and higher concurrent wind speeds. On the other hand, structures in more sheltered locations (for example, along the north shore of Lake Superior within 300 vertical feet of the lake) may be subject to smaller ice thicknesses and lower concurrent wind speeds. Loads from accreted snow or in-cloud icing may be more severe than those from freezing rain. In particular, in-cloud icing, possibly combined with freezing drizzle, appears to be the most significant icing process in eastern Colorado and New Mexico.

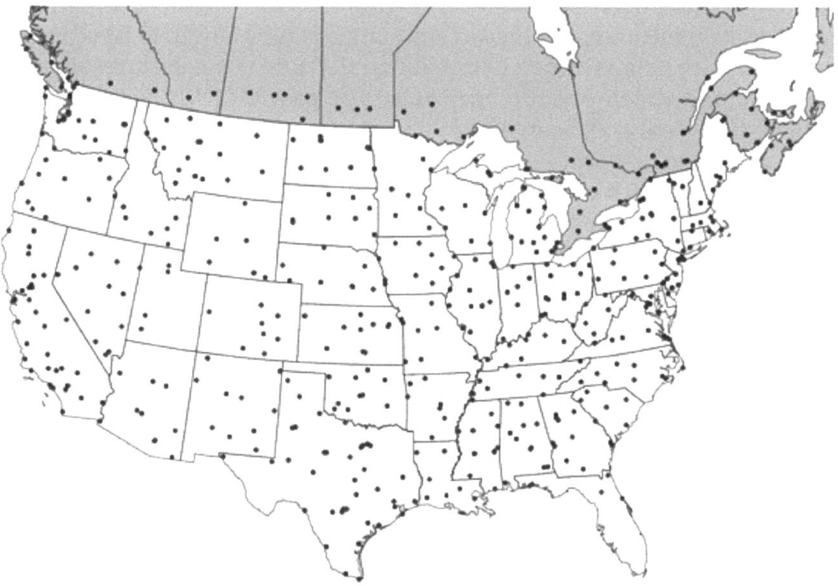

Figure H-1. Locations of weather stations used in preparation of Figures 2-19 through 2-22.
Source: ASCE 7-10 (2010b).

H.4.2 Special Icing Regions

Special icing regions are identified in Figures 2-19 through 2-23. As described above, freezing rain occurs only under special conditions with a cold, relatively thin surface air layer and a layer of warm, moist air aloft. Thus, severe freezing rainstorms at high elevations in mountainous terrain will typically not occur in the same weather systems that cause severe freezing rainstorms at the nearest airport weather station. Furthermore, in these regions ice thicknesses and wind-on-ice loads may vary significantly over short distances because of variations in elevation, topography, and exposure. In these mountainous regions, the values given in Figures 2-19 through 2-23 should be adjusted based on local historical records and experience to account for possibly higher ice loads from both freezing rain and in-cloud icing.

H.5 EXTREME LOADS FROM IN-CLOUD ICING AND STICKY SNOW

Information to produce maps similar to Figures 2-19 through 2-23 for in-cloud icing and snow accretions is not currently available.

H.5.1 In-Cloud Icing

In-cloud icing may cause significant loadings on transmission lines in both mountainous regions and level terrain. In the western United States, in-cloud icing occurs very frequently on exposed ridges and slopes in the mountains. Above the mean freezing level, heavy deposits can form during the numerous storms that strike the region in winter. Steep cliff faces and any exposed structures or obstacles to the wind can become covered with thick coatings of ice. Although in-cloud icing does not commonly occur below elevations of about 3,000 ft (915 m), it does occasionally occur when freezing fog fills the basin regions of eastern Washington and Oregon during periods of strong wintertime temperature inversions. In the eastern plains of Colorado in February 1978, severe rime ice loads were caused by an upslope fog with winds of 10 to 15 mph (4 to 7 m/s). In Arizona, New Mexico, and the panhandles of Texas and Oklahoma, the US Forest Service specifies ice loads due to in-cloud icing for structures constructed at specific mountaintop sites (USFS 1994). In-cloud icing also occurs in the eastern United States, primarily on higher peaks in the Appalachian Mountains. On Mount Washington in New Hampshire [6,280 ft (1,910 m)], the highest peak in the northeastern United States, in-cloud icing occurs about 50% of the time from November through April, with icing episodes typically lasting less than a day and the temperature remaining below freezing between episodes. Typical liquid water contents range from 0.3 to 0.6 g/m^3 and typical wind speeds during icing range from 31 to 62 mph (14 to 28 m/s), with wind speeds greater than 90 mph (40 m/s) occurring 2% of the time. On the more numerous 4,000 ft (1,200 m) mountain summits, in-cloud icing is less severe because the peaks are not exposed to supercooled clouds as frequently and wind speeds are lower. In-cloud icing loads are sensitive to terrain exposure and to the direction of the flow of moisture-laden clouds. Large differences in ice thickness can occur over a few hundred feet distance and can cause severe load unbalances. Advice from a meteorologist familiar with the area is particularly valuable in these circumstances.

H.5.2 Snow

Snow accretions can occur anywhere that snow falls, even in regions that may experience only one or two snowstorms a year. In some regions, extreme accreted snow loads are greater than ice loads from freezing rain or drizzle. A heavy, wet snow storm on March 29, 1976, caused $15 million in damage to the electric transmission and distribution system of Nebraska Public Power District (1976). Mozer and West (1983) reported a transmission line failure on December 2, 1974, near Lonaconing, Maryland, due to heavy, wet snow of 5 in. (127 mm) radial thickness on the wires with an estimated density of 19 pcf (304 kg/m^3). Goodwin et al. (1983) reported measurements of snow accretions on wires in Pennsylvania with an approximate radial thickness of 4 in. (102 mm). The meteorological conditions along a transmission line that failed under vertical load in the Front Range of Colorado were analyzed after the failure. The study indicated that the failure was caused by a 1.7 inch (43 mm) radial thickness, 30 pcf (480 kg/m^3) wet snow accretion with a 42 mph (19 m/s) wind. The return period for this snow load was estimated to be 25 years (McCormick and Pohlman 1993). In the winters of 1994–1995 and 1996–1997, Golden Valley Electric Association in Fairbanks, Alaska, made 27 field measurements of the radial thickness and density of dry snow accretions. Densities ranged from 1.4 to 8 pcf (22 to 128 kg/m^3) and radial thicknesses were up to 4.4 in. (112 mm). The heaviest were equivalent in weight to a 1 in. (25 mm) uniform radial thickness of glaze ice (Golden Valley Electric Association, unpublished data, 1997). Finstad et al. (2009) describes the modeling of sticky snow loads in Alberta, Canada, using weather data.

H.6 OTHER SOURCES OF INFORMATION

Bennett (1959) presents the geographical distribution of the occurrence of ice on utility wires from data compiled by various railroad, electric power, and telephone associations covering the 9-year period from the winter of 1928–1929 through the winter of 1936–1937. The data include measurements of all forms of ice accretion on wires, including glaze ice, rime ice, and accreted snow, but does not differentiate between them. Ice thicknesses were measured on wires of various diameters, heights aboveground, and exposures. No standardized technique was used in measuring the thickness. The maximum ice thickness observed during the 9-year period in each of 975 squares, 60 miles (97 km) on a side, in a grid covering the contiguous United States was reported. In every state except Florida, measurements of accretions, with unknown densities, of approximately 1 inch radial thickness were reported. The map shows measurements as high as 2 inches (51 mm) in the Northeast, Southeast, and South; 1.75 inches (44 mm) in the Midwest; 2.4 inches (61 mm) in the High Plains; and 3 inches (76 mm) in the West. Information on the geographical distribution of the number of storms in this 9-year period with ice accretions greater than specified thicknesses is also included in the Bennett report.

Tattelman and Gringorten (1973) reviewed ice load data, storm descriptions, and damage estimates in several meteorological publications to estimate maximum ice thicknesses with a 50-year return period in each of seven regions in the United States.

In *Storm Data*, storms are sorted by state within each month. The compilation of this qualitative information on storms causing damaging ice accretions in a particular region can be used to estimate the severity of ice and wind-on-ice loads. The Electric Power Research Institute (EPRI) has compiled a database of icing storms from the reports in *Storm Data*. Damage severity maps have also been prepared (Shan and Marr 1996).

Robbins and Cortinas (1996) and Bernstein and Brown (1997) provide information on freezing rain climatology for the 48 contiguous states based on meteorological data. For Alaska, available information indicates that moderate to severe ice loads of all types can be expected. The measurements made by Golden Valley Electric Association are consistent in magnitude with visual observations across a broad area of central Alaska (Peabody 1993). Several meteorological studies using ice accretion models to determine ice loads have been conducted for high-voltage transmission lines in Alaska (Richmond 1985, 1991, 1992; Gouze and Richmond 1982a, b; Peterka et al. 1996). Glaze ice accretions for a 50-year return period range from 0.25 to 1.5 inches (6 to 38 mm) radial thickness, snow from 1.0 to 5.5 inches (25 to 140 mm) radial thickness, and rime from 0.5 to 6.0 inches (12 to 150 mm) radial thickness. The assumed accretion densities were glaze 57 pcf (910 kg/m^3), snow 5 to 31 pcf (80 to 500 kg/m^3), and rime 25 pcf (400 kg/m^3). These ice thicknesses are valid only for the particular regions studied and are highly dependent on the elevation and local terrain features. Large accretions of snow have been observed in most areas of Alaska that have overhead lines.

In areas where little information on ice loads is available, it is recommended that a meteorologist familiar with atmospheric icing be consulted.

It should be noted that taller structures may accrete more ice because of higher winds and colder temperatures aloft, and that the influences of elevation, complex relief, proximity to water, and potential for unbalanced loading are significant.

H.7 CURRENT PRACTICE

A 1979 survey of design practices for transmission line loadings (ASCE 1982) obtained responses from 130 utilities operating 290,000 miles (470,000 km) of high-voltage transmission lines. Fifty-eight of these utilities specifically indicated "heavy icing areas" as one reason for special loadings in excess of NESC requirements. Design ice loads on conductors ranged from no ice (primarily in portions of the southern United States), up to a 2 or 2.25 inch (50 or 57 mm) radial thickness of glaze ice in some states. Radial glaze ice thicknesses between 1.25 and 1.75 inches (32 and 45 mm) are commonly used. Most of the responding utilities design for heavy ice on the wires with no wind and less ice with wind. Few utilities consider ice on the supporting structures in design.

The *National Electrical Safety Code* (IEEE 2017) defines four geographical loading districts that specify combined ice and wind loads. A discussion is provided in the handbook published with the 3rd edition of the NESC (National Bureau of Standards 1920):

> The assumed ice loadings have been chosen after careful consideration of data obtained from the U.S. Weather Bureau, from electric companies, and from engineers. The values chosen do not represent the most severe cases recorded, but do represent conditions that occur more or less frequently. Ice loading of 1/2 inch is frequently exceeded, particularly near the northern and eastern borders of the U.S., and on occasions ice has been known to collect to a thickness of 1.5 inches and even more.

In addition to NESC loading districts (Heavy, Medium, Light, and Warm Island), the NESC has adopted the 50-year glaze ice map in the previous editions of this manual (ASCE 1991, 2010a) to establish ice loading criteria.

APPENDIX I
SUPPLEMENTAL INFORMATION REGARDING LONGITUDINAL LOADS

I.1 INTRODUCTION

Longitudinal loads are complex and may be created through many different events including differential intact wire tensions, broken wires, and construction loads. When considering longitudinal loads, dead-end structures are, by definition, capable of supporting full wire tension loads with all wires removed in one longitudinal direction. Strain, angle, and tangent structures are typically designed for longitudinal loads much less than full (one-side only) wire loads. This appendix will focus primarily on longitudinal loading criteria, calculation methods, and failure containment approaches for tangent, angle, and strain structures. A discussion on transverse cascades is also included.

Longitudinal loads may exceed those imposed by the intact wire system. The potential for extreme longitudinal loads necessitates that utility owners consider including longitudinal loads as part of their standard structural loading criteria. Due to the diversity in regions, reliability needs, and structure types, the approach to longitudinal loading criteria taken by utility owners varies greatly. Longitudinal loading criteria should be collaboratively developed by key stakeholders that have an influence on the reliability of a transmission line. The concepts in this manual have been published in the industry and should be considered for an owner's longitudinal loading criteria. It should be noted that longitudinal loading is a very complex issue and the calculation and failure containment approaches are not limited to those presented in this manual.

I.2 LONGITUDINAL LOADS ON INTACT WIRE SYSTEMS

Transmission structures are typically designed to resist the longitudinal imbalances caused by differential tensions in adjacent spans. These differential tensions can be caused by significantly differing adjacent span lengths, topography, and by differing environmental conditions on adjacent spans. One such example would be unequal ice loading. This can result from unequal ice deposits on adjacent spans, or from one span shedding its ice before the ice is shed on the adjacent span.

When calculating the longitudinal loading for a structure, it is important to consider the ability of the wire supports (i.e., insulators) to swing longitudinally, thereby reducing the longitudinal loading on the structure. A typical conductor suspension assembly can move longitudinally, which helps reduce the tension imbalance in the conductors and can reduce the longitudinal loading on the structure. The shorter the suspension assembly, the more the longitudinal loading will be transferred to the structure. Post insulators, depending on their length, end support conditions, and the material used, can have varying degrees of flexibility. Shield wire attachments are generally considered rigid connections, much like strain insulators.

In addition to the flexibility of the wire attachments, the longitudinal flexibility of the supporting structure itself can affect the longitudinal loading experienced by the structure. As the structure deflects due to the load imbalance, the load imbalance is often reduced. Disregarding the flexibility of the structural system (structure and wire attachments) can result in unnecessarily conservative structure design.

I.2.1 Suspension Supports

Unequal wind or ice loads on adjacent spans and conductor temperature variation on unequal adjacent span lengths can result in differential wire tensions. On suspension structures, the resultant structural load due to differential wire tensions is usually reduced by the swing of the suspension insulator strings (Cluts and Angelos 1977). The longitudinal loads transmitted to the structures by the inclined suspension strings rarely exceed 10% to 20% of the conductor bare wire tension, except in hilly or mountainous terrain where in-cloud icing is a hazard. Suggestions for the determination of unbalanced ice loads can be found in IEC 60286 (2003b), EPRI EL-643 (1978), and EN 50341 (CENELEC 2001).

Many lines exist with unequal spans. A single suspension supported short span may not present a problem because it takes very little longitudinal movement at the support to equalize the tensions. However, a span that is significantly longer than the others will attempt to keep the tension balanced at the supports as the wire contracts or elongates. In this case, more longitudinal movement at the support than is required to relieve the tension imbalance may not be available and the imbalance longitudinal load is transferred to the structure. Similarly, one needs to exercise caution where there are several long spans in succession followed by several short spans. The longitudinal imbalance or movement can be quite large at the transitions. Designers may place strain or dead-end structures at the transitions between span lengths.

I.2.1.1 Impact of Slack in Adjacent Spans In the case of in-cloud icing, the longitudinal loads can be significant because ice deposits can vary greatly from span to span (see Chapter 2, Section 2.3.6 and Appendix H). Unloaded adjacent spans with significant slack permit the insulator to swing sufficiently to turn the suspension assembly into a strain support that is likely to transfer nearly all the differential tension to the structure. In this context, slack is defined as the difference between the actual wire length and the straight-line distance between the attachment points. Problems have been observed in areas where the slack difference of adjacent spans exceeds twice the length of the insulator strings.

Assuming level spans and using the parabolic approximation of the catenary, the following relationships may be used:

$$\text{sag} \cong \frac{wS^2}{8T_H} \tag{I-1}$$

$$\text{slack} \cong 8\frac{\text{sag}^2}{3S} = \frac{S^3}{24\left(\frac{T_H}{w}\right)^2} \tag{I-2}$$

where

w = Wire unit weight,
S = Straight-line span length, and
T_H = Horizontal component of tension.

When the conductor within a ruling span section is not longitudinally restrained (generally true at suspension insulators), slack is proportional to the cube of the ruling span. For example, using a T_H/w ratio of 5,000 ft, the slack in an 800 ft span equals 0.85 ft. Using the cube relationship, the slack in a span double the length (1,600 ft) would be 8 × 0.85, which equals 6.8 ft.

I.2.1.2 Shield Wire Supports Shield wire supports often have short suspension linkages that provide insignificant reduction of unbalanced wire loads. Differential tensions may develop at shield wire supports during significant changes (low to high) in ambient temperatures. Generally, designers will specify an appropriate longitudinal load to account for these changes in temperature.

If the line is located in an in-cloud icing area, the shield wire support of suspension structures may pose the highest risk for structure failure because the differential shield wire tensions could be significantly higher than the differential conductor tensions produced by the same weather conditions. Designers have utilized several methods to reduce the differential shield wire tensions caused by in-cloud icing. Examples include longer suspension links to provide more flexibility to reduce tension imbalance; slip or release clamps to limit the maximum load acting on a support point; and, in some cases, shield wire supports have been designed to act as fuses to collapse at defined loads, thereby preventing more serious damage. Some utilities have removed the shield wires from lines located in areas likely to experience in-cloud icing. However, if removing the shield wire, the designer should address the potential impact on grounding and lightning protection.

I.2.2 Strain Supports

Strain supports by definition are designed to resist differential tensions (longitudinal loads) from adjacent spans, but they may be designed for less than full dead-end capability.

I.3 LONGITUDINAL LOAD CALCULATIONS

If the failure of a component, such as a broken wire, could cause a cascading failure of successive tangent structures not designed to resist wire tension loads, failure-related load criteria should be considered to minimize the extent of the damage and the time required to restore service (EPRI 1979). Failure-related load criteria, such as the broken wire load (BWL), have been used successfully to mitigate the effects of severe differential wire tensions and to minimize the extent of a failure. Based on experience (Ostendorp 1997), the breakage of conductors, shield wires, components, and line cascades are serious problems. The potential for a cascade exists when sufficient slack is introduced into a span so that the unbalanced longitudinal load at the adjacent structure is significant enough that it could fail that structure. As the second structure fails to resist the residual load, it allows the wire to move on to the next structure and repeat the sequence, resulting in a cascading failure. To stop the cascading, it is necessary to limit slack transfer.

While the dynamic loads immediately after a phase or wire break could cause instantaneous loads to be higher than the static intact condition, it is typical industry practice not to include these dynamic loads in the design of structures. Dynamic loads are localized, impact only the adjacent three structures or less, and quickly dissipate.

Most structure types can be designed to provide some longitudinal strength resulting in increased resistance to cascading failure for a small increase in initial cost. Two common, simple methods are often used to estimate an unbalanced longitudinal load: the Residual Static Load (RSL) Method and the EPRI Method.

It should be noted that the unbalanced longitudinal loads determined using RSL and EPRI methods constitute the minimum required "static" loads to be resisted by the structures to avoid cascading failures. Note: The RSL factors do not consider dynamic effects.) The calculated unbalanced longitudinal loads act on the support structure in the direction away from the initiating failure event and should be considered to act concurrently with the effects of any permanently applied load imbalance.

I.3.1 Residual Static Load Method

The RSL Method determines a longitudinal load factor that is applied to the wire tension in order to determine the *residual static load* (RSL) on structures after a wire failure and after all dynamic effects from the wire break have diminished. The reduction in the load magnitude resulting from the insulator swing and support deflection may be considered in the calculation of the RSL. Computer programs (EPRI 1983, Mozer et al. 1977, Peyrot 1985) and design charts (Comellini and Manuzio 1968, EPRI 1978) can be used to assist in the calculation of the RSL values. RSL values derived from the Comellini and Manuzio charts are based on insulator string length and span length and will provide values of approximately 60% to 70% of everyday tensions.

Figure I-1 provides RSL longitudinal load factors as a function of the span/sag ratio and the span/insulator ratio. The calculation assumes rigid supports (i.e., the potential benefiting effects of the flexibility of the supports are neglected) and 10 equal-length spans between the wire break and the next dead end. The span/insulator ratio is the ratio of the average span length within a given tension section to the average effective insulator length (i.e., the insulator length free to swing longitudinally).

The RSL is calculated for the bare wire (no ice or wind) loading condition at an average temperature. The RSLs are applied to one of the conductor support points or to one (or both) shield wire support point(s) on a structure. RSLs are applied in only one longitudinal direction, along with 50% or more of the intact wire vertical load. The other support points on the structure will be considered in the intact condition.

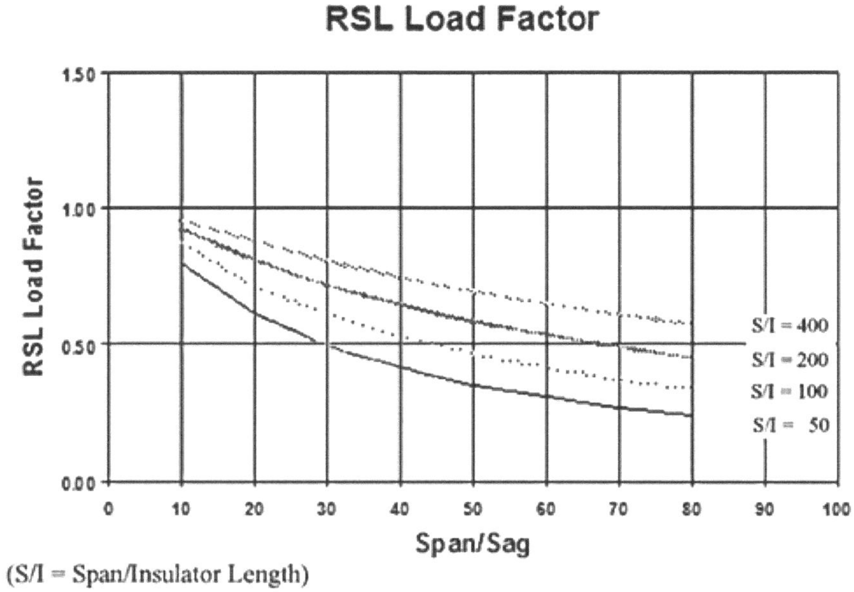

Figure I-1. RSL Method longitudinal load factor.

I.3.2 Electric Power Research Institute Method

The Electric Power Research Institute (EPRI) developed another methodology for calculation of unbalanced longitudinal loads. This method was developed from research completed by EPRI, and calculates loads as a function of the horizontal wire tension, the span/sag ratio, the span/insulator ratio, and the support flexibility (Ostendorp 1997). Although this method can also approximately predict the impact load on the structures adjacent to the initial failure, due to the complexity of those calculations, that portion of the EPRI method is not presented here.

Figure I-2 provides longitudinal load factors as a function of the span/sag ratio and the stiffness of the support structures. Wire tensions multiplied by the longitudinal load factors provide approximate design loads that include dynamic effects, structural stiffness, and insulator lengths. The span/sag ratio is the ratio of the average span length within a given tension section to the sag of the average span for a given conductor or shield wire tension. Longitudinal load factors are provided for "rigid" structures such as guyed or latticed structures of high stiffness, as well as for "flexible" structures such as single poles capable of enduring large elastic deformations.

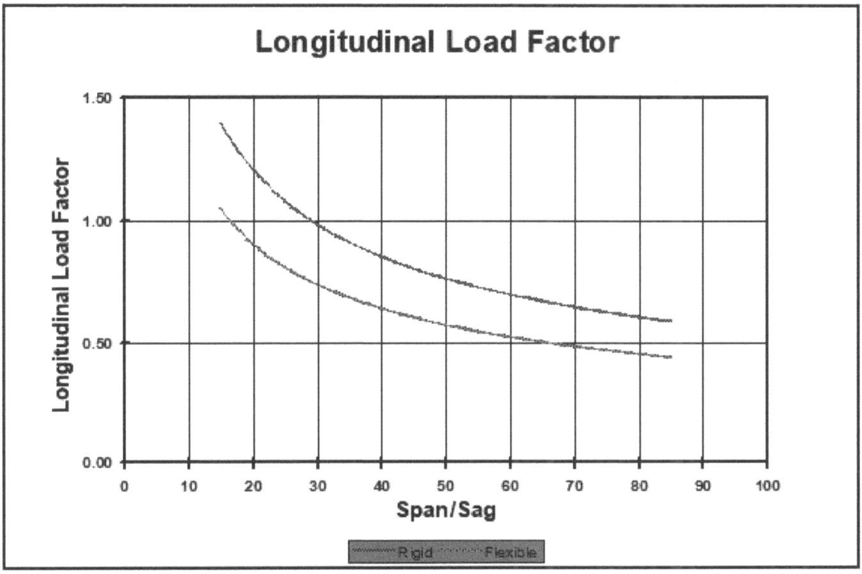

Figure I-2. EPRI Method longitudinal load factor.

I.4 FAILURE CONTAINMENT APPROACHES

Transmission lines may be exposed to severe wind and ice loads, vehicular impact, and other extreme events that may result in structural loading exceeding the criteria for which the transmission structures were designed. Additionally, if a structural or hardware failure occurs, longitudinal loading may exceed design structural loading on several adjacent structures. When longitudinal loading exceeds design structural loading, infrequent failures of a few structures or components due to these extreme events is a generally accepted practice among utilities and utility maintenance practices should be planned accordingly. However, it is recommended that the designer of the line consider developing structural loading criteria that are coordinated with the maintenance practices of the utility. The structural loading criteria should also provide the utility with the desired transmission line performance under extreme loading so that transmission line failures are contained as desired.

When selecting cascade failure mitigation methods, the designer should consider several factors:

- The inherent longitudinal strength of the structure types being used: Square-based latticed towers are inherently stronger in the longitudinal direction than H-frame structures. Including a modest longitudinal design load may not affect the cost of the latticed towers, while it may be too costly or not feasible for a self-supported H-frame structure.

- Criticality of the transmission line: Lines that do not have any redundancy or have such high demand that an extended, unplanned outage would cause issues elsewhere in the system are often considered critical assets. As such, increased up-front spending to significantly limit the extent of any potential failure may be warranted.
- Degree of difficulty in restoring the line: Some lines are more difficult to restore than others. There may be significant topography or environmentally sensitive areas that make access difficult. Some structures on the line may be unique and difficult to replace, such as extra-tall structures at a crossing. The subsurface conditions may make locating or constructing foundations difficult (e.g., very hard rock, karstic formations, significant below-grade infrastructure).
- Cost and availability of replacement materials: If a line can be easily restored with materials that are readily available, it may be less critical to limit the extent of a failure. However, if a line could only be restored using expensive or long-lead materials, then it might make more sense to pursue increased failure containment measures.
- Construction cost differential for each method: Some structures have inherent longitudinal strength and as such can be designed for failure containment loads with little or no increase in costs. In cases where designing each structure for failure containment loads would not be economical, it may make more sense to install periodic stop structures.

Successful failure containment may be achieved by providing sufficient longitudinal strength (1) on all structures, or (2) on failure containment structures inserted at regular intervals. Angle structures may be used as failure containment structures if their longitudinal strength is sufficient to resist the unbalanced loads and to arrest a cascading failure.

I.4.1 Failure Containment Philosophy

Initial transmission line failures may be caused by several different types of events such as train derailment, a major tornado, a low-flying aircraft, or a severe ice storm. An event may bring several structures to the ground and may be accompanied by component and wire failures. The failures could create complicated dynamic forces at adjacent structures. The arduous effort needed to quantify the dynamic energy or impact component at the adjacent structures has directed attention to the security (or survival) of the second, third, fourth, or fifth structure away from the initial failure (Thomas 1981, Ostendorp 1997, Kempner 1997).

Depending on the importance of the line, it is generally agreed that if the second, third, fourth, or fifth structure from the initiating event does not fail, there will be no cascade and most of the energy released by the failure will have dissipated. Therefore, the problem of failure containment may be reduced to the problem of determining the required longitudinal strength to resist the differential tensions at the second, third, fourth, or fifth structure, respectively, while allowing the failure of one or more structures to dissipate the released energy.

I.4.2 Basic Assumptions

It should be noted that any event that permits the creation of excessive slack is likely to produce longitudinal loads that may lead to a cascading failure. Longitudinal cascades of high-voltage lines (Frandsen and Juul 1976) have resulted from initial failures other than broken shield wires or conductors. For example, failure of a heavy angle structure could introduce excessive conductor slack and longitudinal loads that trigger cascading failures on both sides of the fallen angle structure.

I.4.3 Failure Containment Approaches

H-frames and narrow-based, rectangular, latticed structures have little inherent ability to withstand the longitudinal loads of a cascading line. Additionally, the shield wires attached to these structures with near-rigid attachments may contribute to or initiate a cascade. It is considered prudent design practice to employ methods to limit the length of a cascade. For existing lines with limited longitudinal strength, the cost of strengthening such structures or adding longitudinal guys to tangent supports is likely to be prohibitive, undesirable, or ineffective. Another option would be to insert failure containment structures (e.g., stop structures, anchor structures, anti-cascading structures, full dead-ends) at prescribed intervals along the line to limit the extent of the damage caused by a component, structure, or foundation failure.

I.4.3.1 Failure Containment Structures Design codes, criteria, and philosophies for failure containment differ by regions, transmission line owners, and system operators. A typical distance interval between failure containment structures varies but may be as long as 16.1 km. Judgment of these distances may include the length and importance of the line, longitudinal strength of the suspension structures, terrain, land use, restoration time, emergency stocking levels, cost, right-of-way access, and proximity to facilities that may be affected, such as railways, interstate roadways, etc. The decision to install or create anti-cascade or stop towers at intervals along an existing line requires an awareness of the means by which a longitudinal cascade is propagated. Failure to appreciate the mechanics involved may negate the entire effort.

One consideration that should not be overlooked in the placement of anti-cascading structures is their role in the construction stringing process. Anti-cascading structures could be located with input from experienced construction personnel to incorporate stringing staging, segment construction, and stringing length limitations. This can allow the designer to include anti-cascading structures at the aforementioned intervals and provide construction crews more optimal structure configurations for stringing.

Special resistance structures are typically latticed towers or guyed or self-supported pole structures that provide a sufficient level of strength to resist the unbalanced longitudinal loads caused by the failure of components, wires, or structures. Special resistance structures have traditionally included a structure capable of supporting wires only on one side or with broken wire capability. In general tangent (suspension) structures do not possess sufficient strength to provide resistance to cascading failures. The cost associated with developing such capacity in these suspension structures may be prohibitive.

It is important to note that for high voltage (HV) and extra high voltage (EHV) lines, the attempt to use a rigid suspension structure as a stop tower will not succeed even if the tower itself has great longitudinal strength. Allowing the wire movement to pass on down the line will ensure continuance of the cascade even though the stop towers may remain standing. In such instances, it is recommended that special resistance structures be provided at selected intervals along the line to limit the length of a cascading failure to an acceptable number of structures. A strain-type structure could stop a cascade if it has sufficient strength to resist the unbalanced loads (bare wire or iced, as required) and prevent the movement of wire along the line. A suitably strong strain-type angle tower will serve this purpose. In new line construction, the frequent need for angle structures may be accepted as a design alternative to building the anti-cascade strength into each suspension structure.

On low-voltage H-frames or portal structures, the length of the insulator strings will approximately equal the available deflection; therefore, application of longitudinal storm guys, and possibly the installment of metal crossarms at appropriate intervals, can be a rational means of removing the threat of long cascades. Such applications at major highway or railroad crossings should consider the impact of broken wire loading on the behavior of containment structures. At more general locations, RSL may be specified because the impact is diminished to a residual tension after a number of structures a distance away from the break.

The alternative of either inserting or converting a suspension-type structure to perform the anti-cascading duty is attractive but not always possible if the suspension strings are long, as with high voltage (HV) and extra high voltage (EHV) lines. It is possible that enough wire movement will be passed on through the stop tower so the failures continue. The swing of the insulator string will produce a longitudinal load equal to the vertical load being supported at the point, multiplied by the tangent of the angle of swing of the insulator string. This secondary effect must be checked.

I.4.3.2 Install Load Release Mechanisms Release mechanisms that reduce the tension or drop the conductor have been developed but are not widely accepted or used (Peabody 2004). These devices can be divided into two types: (1) Energy-absorbing mechanisms which reduce impact loading and add slack into the span, thereby reducing the RSL; and (2) load release mechanisms which allow the wire to slip through the suspension clamp.

The performance of release mechanisms should be calibrated and verified in representative tests. It is imperative that the design of the slip or release mechanism perform adequately under the design climatic and operational conditions. Regardless of the release device deployed, the structure and supporting hardware must be able to withstand the RSL that the release mechanism does not shed.

Some release mechanisms may not be suitable for use in areas where heavy ice buildups are frequent. Premature release of the device under unbalanced ice could result in a dangerous and undetected ground clearance violation that may constitute a danger to the public.

I.4.3.3 Limited Structure Failure Design Transmission structures can be designed to have a specific component failure mode so that a structural component such as the arm will have less longitudinal capacity than the structure body (Bryant 2012). In this case, the longitudinal capacity of the arm would be designed so that it would fail prior to other structure components. This would allow the longitudinal load to balance and/or dissipate while the core structure body remains undamaged. This can lead to a decrease in recovery time after a failure event by limiting the structure damage to certain components. Designers should also consider the durability, mode, and rate of local component failure. In general, plastic failure is considered to be preferable to shear or tension failure in mechanical fuses that may fail rapidly and induce a dynamic load into the system.

I.4.3.4 Failure Containment for Icing Events In areas where icing events are frequent, utilities may adopt failure containment loads with iced conductors as a design requirement for important lines. Specially reinforced or guyed structures may be used at regular intervals to resist the extremely large differential tensions. These same structures may also be capable of arresting a cascading failure.

I.4.3.5 Failure Containment: Bonneville Power Administration Method A system approach can be followed to mitigate the effects of failure-related unbalanced longitudinal loads on transmission lines. The system approach (Kempner 1997) uses a "failure containment" philosophy that accepts the failure of one tower on each side of the initiating event. The longitudinal loading case assumptions are (1) only one wire or phase is broken at one time, and (2) the break occurs during an everyday load situation, which is defined as no ice, no wind, a conductor temperature of 30 °F (−1.1 °C), and initial sag. The conductor tension obtained under these conditions is multiplied by an impact factor. Standard suspension towers (0° to 3° line angle) and "heavy" suspension towers (0° to 6° line angle) have an impact factor of 1.33. The impact factor for "light" suspension towers (no line angle) is 0.67.

Suspension Tower Conductor The load case consists of

- A vertical load at the broken conductor attachment point [i.e., 50% of the conductor weight and hardware at 30 °F (−1.1 °C)] and the vertical load at the attachment point of intact conductors (i.e., the weight of the conductors and hardware).
- A longitudinal load at the support (i.e., bare wire everyday tension multiplied by the appropriate impact factor).
- A transverse load caused by line angle. Only one phase is assumed broken for both single- and double-circuit towers. Each conductor attachment point shall be considered individually. For a double-circuit tower, this load case shall be repeated with only one circuit strung.

Strain Dead-End Conductor The load case consists of

- A vertical load (i.e., weight of the conductor and hardware) at 0 °F (−17.8 °C).
- A transverse wind load on the tower and wires [i.e., at 40 mph (18 m/s)] with no ice.
- A longitudinal load equal to 125% of sagging tension. The vertical, transverse, and longitudinal wire load is multiplied by a 1.5 load factor. For double-circuit towers, this load case shall be repeated with only one circuit strung.

Shield Wire The load case consists of

- A vertical load of the iced shield wire (i.e., weight of glaze ice equivalent to 1.5 times the working load shield wire design ice thickness at maximum working tension). The equivalent glaze ice thicknesses are light suspension tower, 0.75; standard suspension tower, 1.125; and heavy suspension tower, 1.125. The vertical load is the sum of one-half the equivalent iced wire weight for 1.5 times the transverse span plus one-half the bare weight of 0.5 times the transverse span. Additionally, a vertical conductor load equal to the equivalent ice-coated wire weight is applied to 1.0 times the transverse span.
- The longitudinal load of the shield wire equals the horizontal tension and is applied to all shield wire peaks.

I.4.3.6 Percent of Everyday Wire Tension A design longitudinal load, historically known as broken wire load (BWL) (ASCE 1991), can also be used. Experience has shown that flat or horizontal configurations, single-circuit lines designed with the BWL concept have produced transmission lines with a sufficient level of longitudinal strength to contain the effects of broken wires and other comparable failures that may have otherwise resulted in a cascade.

A horizontal load that is equal to the everyday bare wire tension (EDT) of the shield wire and is equal to about 70% of the EDT of a conductor, applied as a single load at any one support point, has been used successfully to mitigate the effects of broken wires. It should be noted that frequently occurring heavy ice conditions or stiff, brittle supports may require a larger longitudinal load.

I.5 TRANSVERSE CASCADES

I.5.1 Characteristics of a Transverse Cascade

Transverse cascades are differentiated from longitudinal cascades in that the tension, magnitude, and direction of the load of the wire system after the collapse of the initiating structure failure are predominately in the transverse direction. Successive structural failures in a transverse cascade collapse in a generally transverse direction; as such, they may be incorrectly considered a failure caused by a broad front wind. A plan view (top view) of a typical transverse cascade is illustrated in Figure I-3.

Transverse cascades can occur in various ways or from different causes. Examples include (1) HIWs that overload one structure transversely and

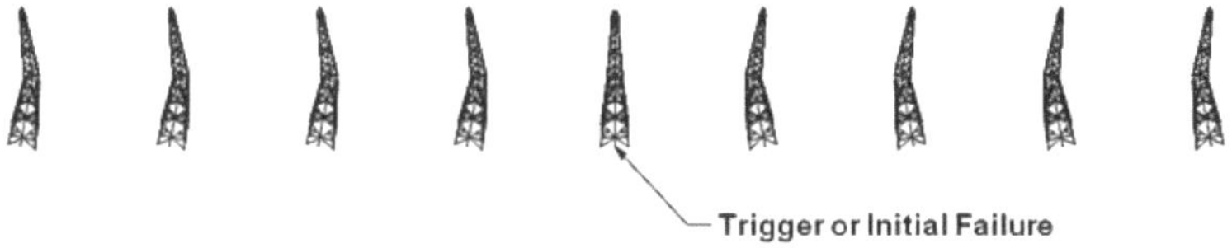

Figure I-3. Plan view of typical transverse cascade.

its failure successively overloads adjacent structures, and (2) on short spans of wood poles, a weak or decayed pole may fail due to transverse wind, which adds wire tension load to the wind load on adjacent structures that then also fail transversely.

Most transverse cascades are initiated by the impact of a high-intensity wind (HIW) on the line, with one or two structures brought down. These small, local failures frequently become transverse cascades of dozens of structures. These failure scenarios have often been misjudged as multiple failures caused by a "wall of wind" overcoming all the fallen structures, when the actual failure mechanism was an initial failure of a single structure due to HIW and subsequently a transverse cascade. Failure of many towers from widespread transverse wind is not common except in areas subject to cyclones, hurricanes, or seaside gales.

An understanding of the loads generated in the wire system after the transverse collapse of one or two structures from HIWs along with an awareness of the line systems (wires and structures) that are vulnerable to these loads can assist the line engineer in providing design options that will reduce the possibility of these types of failures.

I.5.2 Wire Behavior of a Transverse Cascade

A significant parameter in what follows is the slack, which is the difference in length between the straight line joining the points of support and the length of the suspended wire. This exercise uses a parabolic equation because the added precision of working with catenaries is not required. See Equations (I-1) and (I-2). In these two equations, sag is a function of span^2 and slack is a function of span^3.

The ratio of T_H/w is generally referred to as the catenary or parabolic constant. For typical spans with the parabolic constant of tension/unit weight of 487.7 m can be found in Table I-1.

It may be noted that the conductors are supported on suspension insulator strings permitting restricted longitudinal swing varying with the length of the string, whereas the shield wires are almost always firmly attached to the tops of the shield wire peaks of the structures.

Table I-1. Typical Span Characteristics

Span	Sag	Slack
400 ft (121.9 m)	12.5 ft (3.81 m)	1.04 ft (0.317 m)
800 ft (243.8 m)	50.0 ft (15.24 m)	8.33 ft (2.54 m)
1,200 ft (365.8 m)	112.5 ft (34.29 m)	28.13 ft (8.57 m)
1,600 ft (487.7 m)	200.0 ft (60.96 m)	66.67 ft (20.32 m)

With the transverse failure of a single structure, the added length to the wire system can be calculated as well as the transverse and longitudinal loads applied to the adjacent structures. The insulator strings will swing toward the fallen structure, pulling slack from adjacent spans of conductors. However, with the resistance offered by the inclined insulator strings, there will be a great increase in all conductor tensions. These tensions exert longitudinal forces on these towers as well as significant transverse loads.

Shield wire tensions will increase more rapidly with no relief due to insulator string swing, and the pulls exerted on the tops of the shield wire peaks will be limited only by the slip strength of the clamps or the fusing capacity of the shield wire peak itself.

These loads can overwhelm the adjacent two structures, leading to a compression buckling of the mast or nearest corner leg of a latticed structure. However, as the structure starts to fall, the inward tensions start to relax while the tensions back to the next set of adjacent structures will increase. The falling structures will therefore describe an arc in falling, pulled first toward the failed structure but then away from it. Crossarms will strike the ground slightly away from the trigger structure, sometimes as much as 3 ft (1 m).

This pattern of structures falling slightly away from the trigger structure can be readily discerned on site if the investigator is aware of the phenomenon. If the structures on the ground almost "point" back toward the trigger tower and there is further evidence of the failed and outwardly splayed corner legs of a latticed structure, the sequence of events can be confirmed.

I.5.3 Conditions Leading to Transverse Cascading

By examining the parameters that will create the greatest diagonal pulls on adjacent structures, on the one hand it can be noted that

- Short spans contain little slack to relieve the high tensions produced by the falling structure. Short spans also create the greatest tension increases after the failure of one structure.

- Tall structures (such as double-circuit vertical configurations), in falling transversely, lead to large increases in wire loads of upper conductor phases and of the shield wires.
- The short insulator strings of low-voltage lines restrict movement of slack from adjacent spans.

On the other hand, EHV lines are inherently safer with regard to transverse cascading for several reasons:

- Longer insulator strings permit greater equalization or reduction of conductor tensions.
- Strength requirements for carrying the bundled conductors of an EHV line minimize the influence of the shield wire system that usually is similar to that used for lower voltages.
- The longer spans usually associated with EHV also contain larger amounts of slack and do not tighten as quickly when one structure falls.

It should be noted that the reduced influence of the shield wire system on EHV lines may be threatened by the increasing trend of replacing conventional small steel or aluminum-clad steel wire stranding with much larger, heavier, and stronger optical ground wire (OPGW). Replacement with OPGW may require a corresponding strengthening of the clamping and the shield wire peaks themselves.

The shield wire system can, and in most cases does, contribute a major part of the cascade-inducing forces because it is the highest part of the wire system, and the direct clamp system permits no equalization or reduction of tension. A stronger shield wire system, comprised of wires, clamps, and shield wire peaks, increases the potential for a transverse cascade.

APPENDIX J
INVESTIGATION OF TRANSMISSION LINE FAILURES

J.1 INTRODUCTION

Line failures provide a unique and highly valuable opportunity to increase our understanding of transmission line behavior. Not all damage or failures can be avoided, and it is anticipated that failures will occur under extreme conditions that exceed the code-required and utility-established design criteria. A systematic investigation can provide information that may be used to reaffirm or improve design criteria and maintenance practices. The investigation may reveal that the conditions were in excess of design criteria and no modification of the criteria or maintenance practices is justified. The goal of the failure investigation is to establish the cause of the failure and try to reconstruct or understand the behavior of the line subsequent to the failure initiation.

There has been much public reporting of failures in recent years, but little has been published dealing with the technical aspects of transmission line failures. Information on structural failure investigations may be found in publications by Carper (1986) and Janney (1979), and in 1973 a series of papers on transmission line failures (Griffing and Leavengood 1973) was published.

The correct interpretation of the causes of transmission line failures has, at times, led to significant modifications of line design practices. The investigator should be certain that the assumed failure mechanism is consistent with the evidence.

J.2 SAFETY

The site of a transmission line failure can be a deceptively dangerous place, particularly before the restoration and repair crews have secured the area. Although the broken structure may look stable enough, potential danger is all around with the possibility of wires still under tension and/or energized, and structural members balanced in precarious positions.

Figures J-1 and J-2 are two examples of transmission line failures where the site had to be secured prior to initiating the investigation. The cause of failure in Figure J-1 was an extreme wind event. The structure and wires had to be secured before failure investigators could approach the area and proceed with their evaluation. The cause of failure in Figure J-2 was a microburst. This is an obvious example of a failed structure to be avoided until restoration crews are able to secure the site.

At all times, be aware of and follow the applicable utility safety procedures. Document the site from a safe distance until the site is deemed safe to approach either by experience and judgment of the investigator or by the restoration and repair crews.

Figure J-1. Failure of a 230 kV lattice structure.

Figure J-2. Failure of a 46kV wood pole.

J.3 LEGAL CONSIDERATIONS FOR FAILURE INVESTIGATIONS

Fortunately, most transmission line failures do not result in legal actions. However, in the rare event that personal injury, fatalities, or extensive damage to private property are the result of a transmission line failure, the failure investigation team must be aware that more rigorous processes and procedures may be prudent and in some cases required as the data gathered during the failure investigation could potentially be used in legal proceedings.

The investigative team should be aware of and consider the following if it appears that legal action may result from the failure(s) being investigated:

- Defer to all investigative activities of law enforcement, fire, medical, safety, or other governmental agencies. The failure investigation

being conducted on behalf of the utility should not proceed until the site is released.
- Contact legal counsel to advise of the field situation and obtain guidance in fact and data gathering and for contact with parties outside of the utility.
- Take special care when gathering physical evidence. Document its location with photographs prior to moving. Document handling and storage of the physical evidence such that the "chain of custody" is maintained and evidence thoroughly documented.
- Engage a third-party engineering firm or laboratory to perform any testing or analysis deemed necessary.
- In all instances, perform thorough and comprehensive data collection, perform state-of-the-art analysis, rigorously support any conclusions, and behave professionally.

J.4 NEED FOR AND BENEFIT OF THOROUGH INVESTIGATIONS

In any failure event, the utility's responsibility is to ensure public safety and to promptly restore service. Therefore, a predefined emergency response plan should be established so repair and restoration crews can be mobilized quickly and a qualified engineer has adequate time to perform a thorough investigation. Time is the vital factor and, unless plans have been made before the event and priority directives issued, significant evidence and data could be lost.

A utility/transmission line owner should have an established phone list that identifies key failure investigation personnel. These individuals should be familiar with the utility's investigation procedures and policy. The list should be distributed to the utility's line construction and maintenance office(s).

The reasons for attempting to get to the root causes of a failure event are many:

- The cause may be an actual overload of ice, wind, or a combination of the two that exceeded the design specifications and will require an assessment of future risks and costs. The accurate assessment of the actual ice and wind loads is imperative to determine whether there was excessive loading or whether there was a problem or defect within the system.
- Detection of a deficiency or defect may permit modifications to components to prevent further failures or may lead to modifications of current design practices or specifications.

- The cause may be attributed to the deterioration of specific line components that may justify increased inspection and replacement policies.
- Unanticipated dynamic behavior may be detected.
- The investigation may uncover a specific loading case that was not originally considered.
- A systematic and thorough failure investigation should provide the line engineer a greater familiarity with the ways in which the various components of the wire and structural support systems interact when the system is severely stressed.

J.5 CAUSES OF FAILURE

The following lists represent some of the more general causes of transmission line failures.

J.5.1 Natural Phenomena Exceeding Design Criteria

- Extreme wind,
- Extreme ice,
- Combination of ice and wind,
- Landslides,
- Avalanches,
- Ice movement on rivers or lakes (for structures located in the water),
- Flooding (causing damage to structure or to foundation), and
- Soil liquefaction.

J.5.2 Human Causes

- Sabotage, vandalism, or theft of members and/or bolts; and
- Accidental damage caused by equipment and vehicles.

J.5.3 Structure Deficiencies (When Design Criteria Were Not Exceeded)

- Design inadequacies of structure and/or foundation,
- Missing members or loose bolts caused by vibration or omitted during erection,
- Erroneously fabricated members,
- Improperly installed foundations, and
- Deterioration or corrosion of structures.

J.5.4 Conductor, Ground-Wire, and Hardware Deficiencies

- Improper wire splices,
- Faulty or inadequate hardware,
- Fatigue failure of wire or hardware components,
- Insulation failures, and
- Deterioration or corrosion.

J.5.5 Construction-Related Causes

- Excessive vertical load during stringing,
- Excessive longitudinal load during stringing,
- Improper stringing sequence,
- Damage during stringing, and
- Inadequate inspection (of bolted connections, member alignment, pole jacking, method of damping unstrung arms).

J.5.6 Improper Installation of Structures

- Manufacturer's recommended slip joint jacking force was not achieved,
- Field-assembled joints were not inspected prior to stringing,
- Unstrung arms were not dampened or tied off after installation,
- Structure sections were not properly aligned, and
- Manufacturer's bolt tightening procedures were not followed.

J.6 FAILURE INVESTIGATIONS

A failure investigation can be a very simple and quick observation of the facts represented by the evidence. At other times, it will result in a study involving many engineers over a period of years. The least demanding of investigations are those that follow an accident caused by an obvious event, such as aircraft contact, foundation washout, and so forth. The emphasis in the investigation will be directed to finding a means of preventing recurrence and determining whether the post failure behavior of adjacent structures was satisfactory.

A more difficult problem will be encountered when the cause can be identified as a wind or ice storm but the evidence indicates that the failure occurred at lower than the expected design values. These situations require an examination of the evidence to determine whether there was a structure design deficiency. For example, bolts or members may have been missing, foundations may have had inadequate cover, or guy anchors may have had inadequate uplift capacity. In other cases, consideration of yawed or longitudinal wind loads may have been omitted from the design criteria, or probable uplift loads were not considered.

In the case of line damage with multiple failures caused by ice or wind load equal to or exceeding design values, the investigation should attempt to determine the line section that failed by the initial ice and/or wind event. This should be inspected separately from other sections that may have failed due to secondary events. This is an important finding to better understand the behavior of the line, but it is often difficult to distinguish between them.

J.7 PREPARATION

J.7.1 Failure Investigation Equipment

Following is a list of some of the essential items that the investigative team will want to have with them in the field as they conduct their inspection:

- Measuring tape and pocket scale,
- Micrometer (if material sizing is in question),
- Notebook and sketch pad,
- Markers and identification tags,
- Cardboard pieces or 8 × 11 in. paper pad and marker to place in foreground of all photos for future identification,
- Voice recorder,
- High-resolution digital camera with video capability and extra batteries,
- Binoculars,
- Cell phone or radio, and
- Large sealable plastic bags to collect accumulated ice on wires.

J.7.2 Technical Preparation

If time and access permit, the investigators should familiarize themselves with the appropriate line data, conductor and structure loadings, design characteristics, and any special construction records prior to conducting the field investigation. When possible, they should discuss the failure briefly with a group of key design personnel.

J.8 STEPS FOR AN EFFECTIVE FAILURE INVESTIGATION CHECKLIST

The investigators should initiate complete photo documentation, make an overall survey of the damaged area, and listen to viewpoints and evidence of any witnesses or earlier arrivals.

For utilities, first responders typically consist of patrolmen/linemen or switching personnel. In addition to their primary responsibilities during a transmission line failure, first responders are in an opportune position to quickly collect initial data (site photographs are a good example) themselves or from eyewitnesses that may have been on the scene at the time of the failure(s). These reviews are invaluable as the incident is still fresh in everyone's mind. This also requires utilities to train their first responders on such data gathering as well as provide them with time and encouragement to do so.

The following information may be used as a summary checklist for failure investigation but does not cover all the tasks that could be performed during an investigation.

J.8.1 Data Gathering

Field data gathering items could include the following:

- Prior to arriving on site, confirm that the line is locked out (Lock out-Tag out) of service at the substation and de-energized.
- Assume all wires or conductors are live unless confirmed by the local utility. Downed conductors can still have lethal currents. Utility-established practices for work around potentially energized facilities must be followed.
- Develop a plan for conducting the investigation of the failure(s) in the field. Evaluate the nature and extent of the failure(s) and based on the size and available resources of the investigative team(s) develop a plan to deploy to locations that appear to offer the best opportunity to collect the highest quality and most informative data. Communicate with the appropriate operations, maintenance, or construction personnel/organizations to understand the status of the failed facility and restoration plans. Seek out the first responders to the failure for initial impressions and logistics guidance. Provision the investigation team and initiate the field investigation as quickly as practicable.
- The line restoration and repair crews may have already arrived at the site and will be ready to start repair operations. If this happens, try to obtain a visual inspection of the damaged portion of the line. An overall picture taken at this time may provide information and detail that could be lost after the repair activity begins. Observations of marks left by ground impacts may be useful in evaluating the order of failures when secondary failures exist.
- The first impression of the site can result in a multitude of ideas about the failure, and it is valuable for the investigator to record these thoughts.
- Prepare a sketch of the line showing the positions of conductors, insulators, structures, and any indication of the conductors having

been pulled across the ground. Also note the structure configuration, such as the position of the guy anchors, deflected shape of structures, and final position of footing stubs and/or structure legs.

- If the event is an ice storm, attempt to gather representative ice samples from the fallen wires, record the length of each sample along with the diameter of wire that it comes from, and store them in plastic bags for later weighing. Sample ice weights are the best way to accurately measure the ice load that was on the wires.
- An awareness of conductor and shield wire behavior is important because these tie the structures together. Observe and document the position of the wires/insulators in relation to the position of the failed structure.
- If wind is the suspected cause of failure, look for surrounding damage to trees, buildings, and so forth. The Beaufort scale (Baumeister et al. 1978), given in Table J-1, can provide valuable information as to the approximate wind speed.
- Look for signs of the following:
  - Rust on sheared surfaces that may indicate that the bolt or member had loss of cross section prior to the event,
  - Burn marks on the conductor or structure indicating initial point of fault to ground,
  - Evidence of loose or missing bolts, and
  - Shiny steel and worn galvanizing at joints, indicating possible vibration.
- If hardware, insulators, conductors, or overhead ground wires are broken, they may have been triggered by the initial failure or may have been caused by a secondary event. Retrieve and mark some specimens as needed.
- If there are broken wires, note whether the ends of the strands indicate a prior fracture due to fatigue, or a cup cone failure with necking indicative of a tensile failure.
- It may be desirable to remove test sections of steel members for material tests to determine material properties. Record the location of the member samples. Avoid taking samples in the area of high stress because the cold working of the steel will significantly alter its physical properties. If a torch is used to remove the sample, be sure to obtain a sample large enough that a testing coupon can be prepared that has not been degraded due to the localized effects of heat.
- Individuals in the nearby area of the failure may be a possible source of information. These individuals can frequently tell of vibration, galloping, and other unusual meteorological events that may have occurred immediately prior to the failure or historically.
- Data should be maintained in a professional manner. Any data gathered in the field, including photographs, may be discoverable in court.

Table J-1. Beaufort Scale of Wind Intensity

Beaufort number	Wind speed (mph)	Wind effects observed on land	Terms used in US Weather Bureau reports
0	< 1	Calm, smoke rises vertically.	Light
1	1–3	Direction of wind shown by smoke drift but not by wind vanes.	
2	4–7	Wind felt on face, leaves rustle, ordinary vane moved by wind.	
3	8–12	Leaves and small twigs in constant motion, wind extends light flag.	Gentle
4	13–18	Raises dust, loose paper; small branches are moved.	Moderate
5	19–24	Small trees in leaf begin to sway, crested wavelets form on inland waters.	Fresh
6	25–31	Large branches in motion, whistling heard in telegraph wires, umbrellas used with difficulty, wind is heard in buildings.	Strong
7	32–38	Whole trees in motion, difficult walking against wind.	
8	39–46	Branches break off trees, wind generally impedes progress.	
9	47–54	Slight structural damage occurs; chimney pots, slates removed.	Gale
10	55–63	Seldom experienced inland; trees uprooted, considerable structural damage occurs, telephone poles break.	Whole gale
11	64–72	Very rarely experienced, accompanied by widespread damage.	
12	> 73	Very rarely experienced, disastrous damage.	Hurricane

Source: Baumeister et al. (1978), US Weather Bureau.

Figure J-3 is an example of field data collection and the need to contact multiple entities. The cause of the structure failure was a microburst. As part of the failure investigation, first responders from utilities as well as others such as railway officials should be approached for information.

Figure J-3. Failure of a 46 kV wood pole line.

J.8.2 After Returning to the Office

- Obtain weather data from the nearest local weather station.
- Look for additional weather recording stations within the local area of the failure. Examples include substations, telecommunication or cellular stations, highway bridges and overpasses, and homes or businesses.
- Gather design criteria, "as-built" drawings, construction records, and maintenance history of the line.
- When applicable, evaluate failed hardware or send these out for expert evaluation.
- Weigh ice samples or measure the volume of melted ice and calculate the ice load on the wire.

J.8.3 Analysis of Data Gathered

- Look for design inadequacies:
  - Conductor weaker than structure,
  - Combined loading producing critical member stresses not previously considered,
    and
  - Foundation or anchor failures.
- Study field data carefully; try to match field data with postulated cause of failure.
- If the structure appears to have failed below the design load, a more detailed analysis may be warranted, taking into account actual material yield strengths obtained by testing and secondary stresses due to bending and nonlinearities.
- Contact the local utility operations personnel and determine the exact time and nature of the line outage and any other relevant information.
- Examine conductor behavior after the failure event and its potential effect on the remaining transmission line system.
- Ascertain why damage terminated where it did.

J.8.4 Preparation of Report

The report should summarize and document the following:

- Field investigation, observations made, and data collected. The data collected in the field (primarily photographs, sketches, interviews, and notes) should be cataloged for future reference.
- Overview of the physical characteristics and layout of the line, design practices, inspection methods, maintenance practices, and construction techniques prior to the failure.
- Documentation of the failure summarizing the environmental conditions, cause of the failure, identification of initial failure location, sequence of failure, and contributing or mitigating factors.
- Conclusions and recommendations, including adequacy of design criteria, inspection and maintenance practices, effectiveness of failure containment, and recommendations for improvement or modification of new or existing facilities.
- Follow-up evaluation of failure investigation.

J.9 POST-FAILURE BEHAVIOR OR FAILURE CONTAINMENT

The investigation should establish the cause of failure and whether the line performed as designed. If needed, make recommendations regarding:

- Strengthening of the existing structures,
- Improvement of maintenance and inspection procedures,
- Possible change of load adjustments to design criteria and design practices for future lines, and
- Failure containment.

Another function is to identify any evidence of a cascading failure. An initial failure with collapsed structures or broken wires may cause damage to one or two structures adjacent on either side. It is difficult to prevent such damage in all cases because the nature of the initial event and the impact and energy release may not be easily absorbed. If subsequent structures fail, a cascade is more likely. The investigator should also determine the effectiveness of any existing failure containment measures.

J.10 ADDITIONAL REFERENCES: FAILURE INVESTIGATION SPECIFIC

ASCE. 1989. "Guidelines for failure investigation." New York: ASCE Task Committee on Guideline for Failure Investigation, Technical Council on Forensic Engineering.

ASCE. 1997. "Forensic engineering." In *Proc., 1st Congress, Forensic Engineering Division of ASCE*, Minneapolis, Minnesota. New York: ASCE.

ASCE. 2003. *Guidelines for forensic engineering practice*. Reston, VA: ASCE.

EPRI (Electric Power Research Institute). 2003. *The fundamentals of forensic investigation procedures guidebook*, 1001890. Palo Alto, CA: EPRI.

EPRI. 2004. "Forensic analysis of failures." Chapter 5 in *Overhead transmission inspection and assessment guidelines—2004*, 1002007. Palo Alto, CA: EPRI.

EPRI. 2012. "Forensic analysis of failures." Chapter 5 in *Overhead transmission inspection and assessment guidelines—2012*, 10024111. Palo Alto, CA: EPRI.

APPENDIX K
HIGH-INTENSITY WINDS

K.1 INTRODUCTION

Load cases simulating the critical downburst and tornado wind configurations for a generic transmission line system are provided in this appendix.

K.2 DOWNBURSTS

K.2.1 Proposed Critical Downburst Loads

As described in Section 2.2.1, the distribution and magnitude of forces on the line components are a function of the downburst jet velocity V_J, the practical jet diameter D_J, and the location of the downburst wind relative to the target structure, as shown in Figure 2-12. A value of 110 to 160 mph (50 to 70 m/s) is recommended for V_J.

Based on the results of extensive parametric studies conducted using an experimentally validated numerical model, three load cases, providing an envelope for the maximum effect of the downburst wind field on transmission line structures, were suggested by El Damatty and Elawady (2015). One of the load cases requires carrying out nonlinear structural analysis for the wire system. In lieu of nonlinear structural analysis, a set of charts is provided in this appendix to estimate the conductor and shield wire unbalanced longitudinal forces associated with this load case. A list of the nomenclature is provided in Section K.4.

Load Case 1: Transverse Wind Load (Ψ = 0°) This case corresponds to a downburst outflow in the transverse direction. The velocity profile associated with this load case can be described as follows:

- Vertical distribution of the radial velocity along the height of the structure, normalized with respect to the jet velocity V_J, is provided in Figure K-1. An equivalent uniform velocity profile V_{eqt} that results in a base shear and overturning moment equal to or exceeding those obtained from the downburst profile is calculated. It is found that $V_{eqt} = 1.1V_J$.
- Non-uniform distribution of the radial velocity along three spans from each side of the structure of interest is shown in Figure K-2. This symmetric distribution leads to a transverse force acting on the structure of interest. It is found that an equivalent uniform distribution V_{eqc} with a magnitude of $1.06V_J$ can be used to calculate the transverse force on the wires.

Since the 3-second gust velocity is used as the reference velocity for downbursts, a span reduction factor can be applied to simulate the lack of correlation in turbulence that is expected to occur along the conductor's spans. The relation between the conductor span and the reduction factor for the case of downburst was developed by Holmes et al. (2008) and Behncke and Ho (2009). This is compared to the span reduction factor for synoptic wind in Figure K-3.

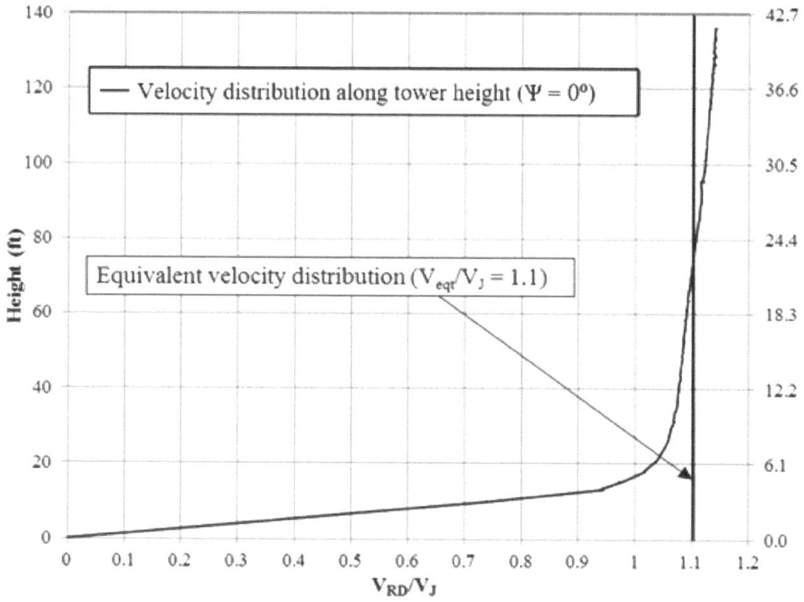

Figure K-1. Radial velocity distribution over structure height for Ψ = 0°. Source: El Damatty and Elawady (2018).

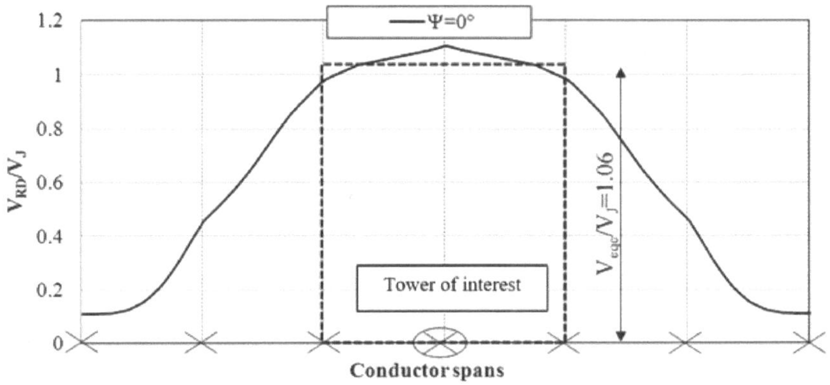

Figure K-2. Transverse radial velocity distribution over six conductor spans for Ψ = 0°. Source: El Damatty and Elawady (2018).

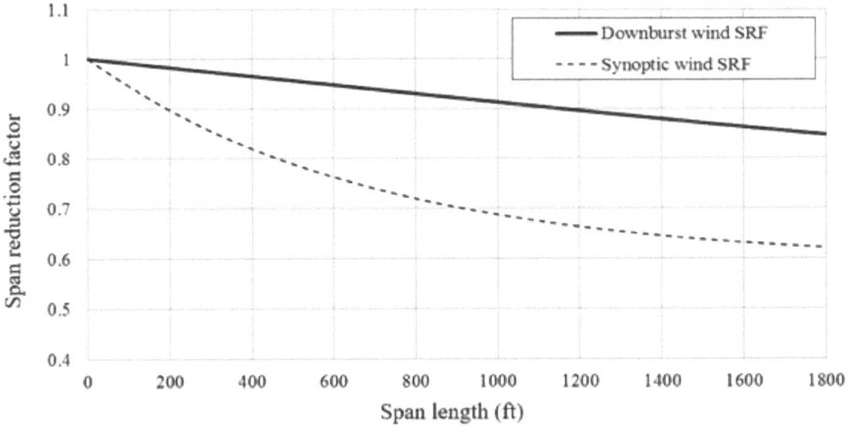

Figure K-3. Downburst span reduction factor for peak pressures as a function of average span length.
Source: Adapted from Holmes et al. (2008) and Behncke and Ho (2009).

Load Case 2: Longitudinal Wind Load (Ψ = 90°) This case corresponds to a downburst outflow in the longitudinal direction. Note that any wires spanning between structures are not loaded in this case. The velocity profile associated with this load case can be described as follows:

- The Vertical distribution of the radial velocity along the height of the structure, normalized with respect to the jet velocity V_J, is shown in Figure K-4. An equivalent uniform velocity profile V_{eqt}

that results in a base shear and overturning moment equal to or exceeding those obtained from the downburst profile is calculated. It is found that $V_{eqt} = 1.1V_J$.

- No force acts on the conductors due to this load case.

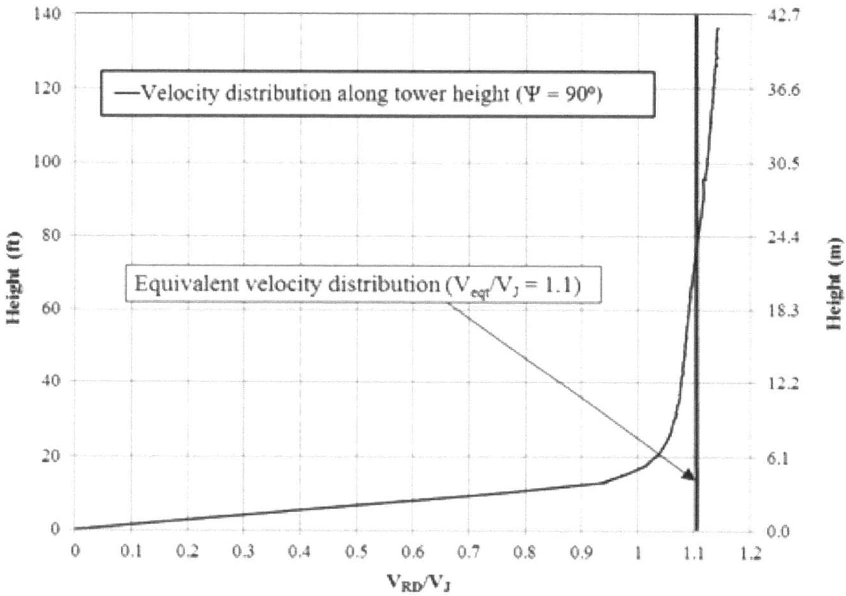

Figure K-4. Longitudinal radial velocity distribution over structure height for Ψ = 90°.
Source: El Damatty and Elawady (2015).

Load Case 3: Oblique Wind Load (Ψ = 30°)

This case is associated with a radial velocity profile that acts at an oblique angle relative to the transmission line. The wind field associated with this configuration will lead to the following effects:

- A velocity profile along the height of the structure in the transverse direction (Y-direction), which can be approximated by an equivalent uniform velocity V_{eqtT} of $0.80V_J$ as shown in Figure K-5.
- A velocity profile along the height of the structure in the longitudinal direction (X-direction), which can be approximated by an equivalent uniform velocity V_{eqtL} of $0.47V_J$ as shown in Figure K-5.
- Velocity profile that acts on the conductors, which has an unequal distribution on the spans adjacent to the structure of interest as shown in Figure K-6. This velocity profile leads to a transverse force, R_Y, as well as a longitudinal unbalanced force, R_X, acting on the structure.

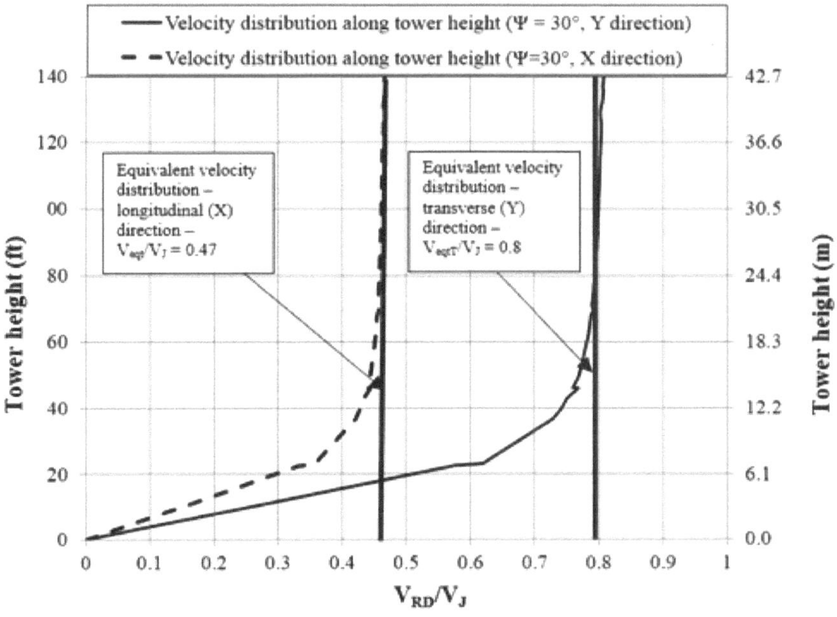

Figure K-5. Radial velocity distribution over structure height for Ψ = 30°. Source: El Damatty and Elawady (2015).

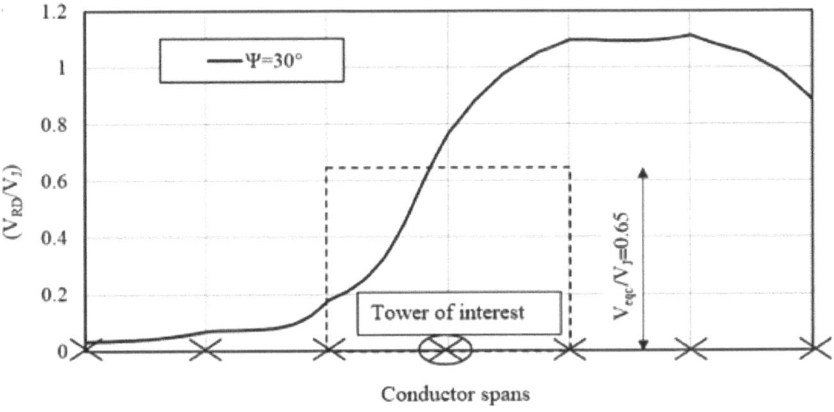

Figure K-6. Radial velocity distribution over six conductor spans for Ψ=30°. Source: El Damatty and Elawady (2015).

An equivalent uniform distribution, V_{eqc}, with a magnitude of approximately $0.65V_J$ can be used to calculate the transverse force on the wires, R_Y, under this load configuration. The evaluation of the longitudinal unbalanced force, R_X, requires a non-linear analysis for the line wires. In order to simplify the analysis, Elawady and El Damatty (2015) developed a set of graphs that are provided in Section K.2.2. These graphs can be used to evaluate the longitudinal unbalanced force of the conductors and shield wires following the procedure described in the next section.

Because the downburst wind speeds are gust wind speeds, the gust response factor, G, and velocity pressure exposure coefficient, K_z, should be considered to be 1.0 for calculating the wind force.

K.2.2 Evaluation of Longitudinal Unbalanced Load R_X for Downburst Case 3

K.2.2.1 Conductor Longitudinal Force Due to the uneven distribution of the loads acting on the conductor spans in this load case, the wire tension forces on the spans adjacent to the target structure are different. This difference is transferred to the structure in the form of unbalanced forces acting in the longitudinal direction of the line. The evaluation of this unbalanced longitudinal load requires performing nonlinear analysis for the conductors, while taking the in-line swing of the insulators into account. The main parameters that affect the magnitude of this longitudinal force are the conductor's projected diameter and weight, span length, conductor sag, the length of suspension insulator, and the downburst equivalent jet velocity (El Damatty et al. 2013).

To simplify these calculations, a set of charts is provided in this appendix from which the longitudinal unbalanced force can be evaluated. Table K-1 shows the upper and lower range of properties considered in the developed charts. In this table, the square of the jet velocity V_J^2 and the wire

Table K-1. Parameters Considered in the Conductor Design Charts

	Lower range		Upper range	
Parameter	SI units	US customary units	SI units	US customary units
Insulator length h	1 m	39.37 in.	5 m	196.85 in.
α	40 m^3/sec^2	0.1244 $mile^3/hr^2$	200 m^3/sec^2	0.6218 $mile^3/hr^2$
Wire weight w	10 N/m	0.6852 lb/ft	40 N/m	2.7408 lb/ft
Wire span L	100 m	328 ft	500 m	1640 ft
Sag ratio (sag/span ratio) S	1 m	39.37 in.	5 m	196.85 in.
Insulator length h	2%	2%	2%	2%

projected diameter d were combined into one parameter, α. This parameter, α, is directly proportional to the transverse loads acting on the conductors. The magnitude of the longitudinal unbalanced force is found to vary linearly with the insulator length h while it varies nonlinearly with the other parameters. However, it is found that by dividing the range of α into four regions and the wire weight w into two regions, the magnitude of the longitudinal forces varies linearly with α, w, and h within each region. As such, for each region, the charts are provided showing the variations of the longitudinal force with the span and sag ratio for the extreme values of the three parameters α, h, and w.

For each of the eight groups (I to VIII), eight graphs are provided for the combinations of the maximum and minimum of the parameter values α, h, and w. Each graph provides the variation of the longitudinal unbalanced force with the span and the sag ratio (Elawady and El Damatty 2015). Linear interpolation can be done between the eight curves to obtain the longitudinal unbalanced force of the considered system as explained in the following steps.

1. Calculate $\alpha = V_J^2 \times d$.
2. Based on the values of α and w, determine to which group the system belongs (I to VIII). Table K-2 provides simple guidance for the selection of the design groups.

Table K-2. Longitudinal Load Chart Guidance

α (m³/sec²)		w (N/m)			
α_{min}	α_{max}	w_{min}	w_{max}	Group	Figure reference
40	80	10	25	I	K-7
40	80	25	40	II	K-8
80	120	10	25	III	K-9
80	120	25	40	IV	K-10
120	160	10	25	V	K-11
120	160	25	40	VI	K-12
160	200	10	25	VII	K-13
160	200	25	40	VIII	K-14

3. Based on the span value and the sag ratio, determine the longitudinal unbalanced force associated with the eight graphs of each group. Those are labeled:
   R_{X1} = Longitudinal force corresponding to (w_{min}, h_{min}, α_{min})
   R_{X2} = Longitudinal force corresponding to (w_{min}, h_{min}, α_{min})

R_{X3} = Longitudinal force corresponding to (w_{min}, h_{min}, α_{min})
R_{X4} = Longitudinal force corresponding to (w_{min}, h_{min}, α_{min})
R_{X5} = Longitudinal force corresponding to (w_{min}, h_{min}, α_{min})
R_{X6} = Longitudinal force corresponding to (w_{min}, h_{min}, α_{min})
R_{X7} = Longitudinal force corresponding to (w_{min}, h_{min}, α_{min})
R_{X8} = Longitudinal force corresponding to (w_{min}, h_{min}, α_{min})
Note: h_{max} = 5 m, w_{max} and α_{max} are the upper range for each group, while h_{min} = 1 m, w_{min} and α_{min} are the lower range for each group.

4. Based on the values of α, h, and w and the above eight evaluated longitudinal forces, linear interpolation can be conducted using this set of equations:

$$R_{X(3-4)} = R_{X3} + (R_{X4} - R_{X3}) \times \frac{(\alpha - \alpha_{min})}{(\alpha_{max} - \alpha_{min})} \quad \text{(K-1)}$$

$$R_{X(5-6)} = R_{X5} + (R_{X6} - R_{X5}) \times \frac{(\alpha - \alpha_{min})}{(\alpha_{max} - \alpha_{min})} \quad \text{(K-2)}$$

$$R_{X(7-8)} = R_{X7} + (R_{X8} - R_{X7}) \times \frac{(\alpha - \alpha_{min})}{(\alpha_{max} - \alpha_{min})} \quad \text{(K-3)}$$

$$R_{X(h_{max})} = R_{X(7-8)} + (R_{X(5-6)} - R_{X(7-8)}) \times \frac{(w_{max} - w)}{(w_{max} - w_{min})} \quad \text{(K-4)}$$

$$R_X = R_{X(h_{max})} + \frac{(R_{X(h_{min})} - R_{X(h_{max})})}{(h_{max} - h_{min})} \times (h_{max} - h) \quad \text{(K-5)}$$

$$R_{X(h_{max})} = R_{X(7-8)} + (R_{X(5-6)} - R_{X(7-8)}) \times \frac{(w_{max} - w)}{(w_{max} - w_{min})} \quad \text{(K-6)}$$

$$R_{X(3-4)} = R_{X3} + (R_{X4} - R_{X3}) \times \frac{(\alpha - \alpha_{min})}{(\alpha_{max} - \alpha_{min})} \quad \text{(K-7)}$$

where R_x is the critical longitudinal unbalanced force resulting from this downburst load case and is to be applied at the insulator to structure connection.

5. The preceding analysis has been performed for a single conductor. For cases of multiple conductor bundles, the longitudinal unbalanced force obtained from Equation (K-7) must be multiplied by the number of subconductors in the bundle.

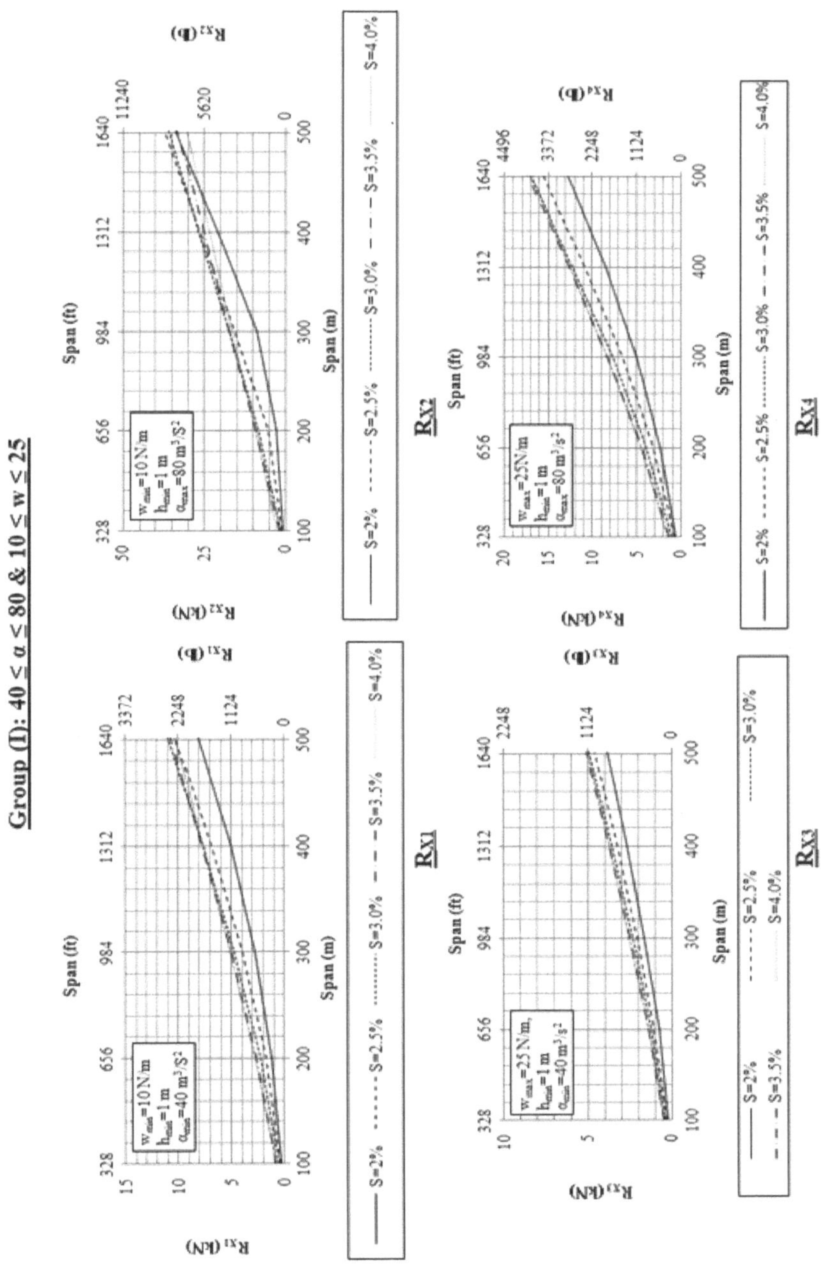

Figure K-7. Longitudinal force (R_X) charts: Group I

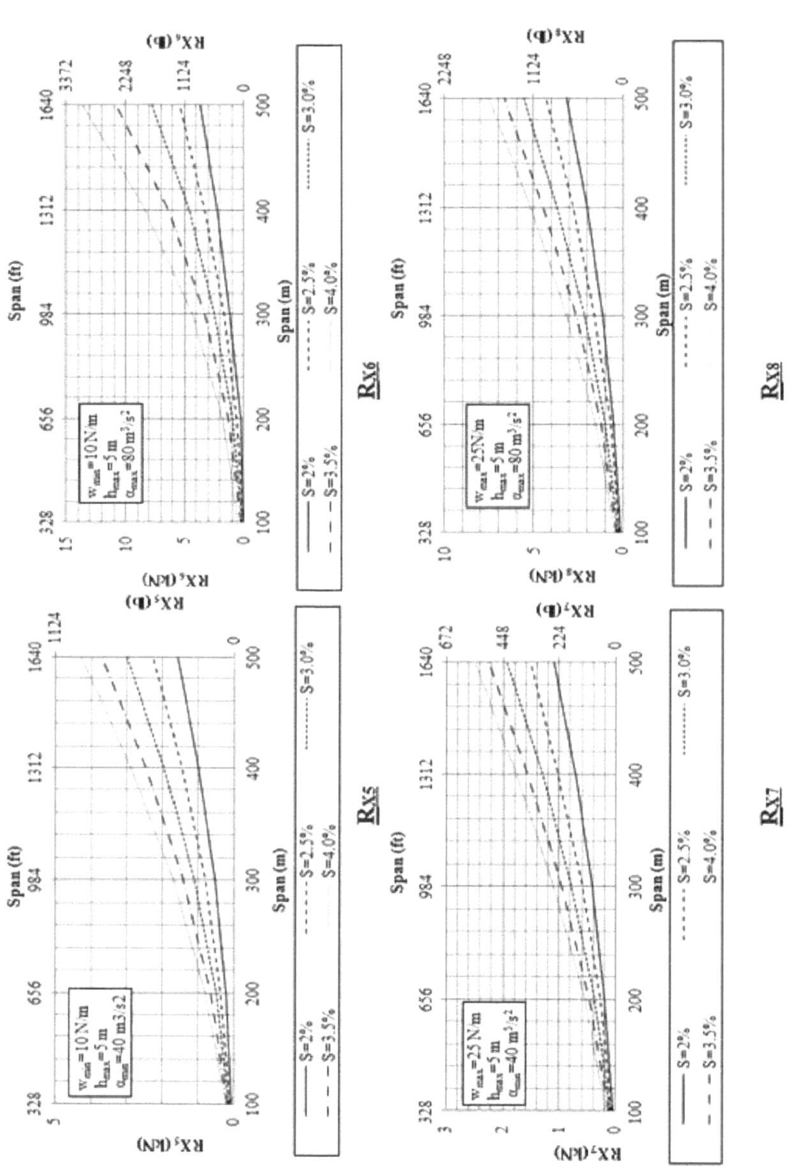

Figure K-7 (Continued). Longitudinal force (R_X) charts: Group I.

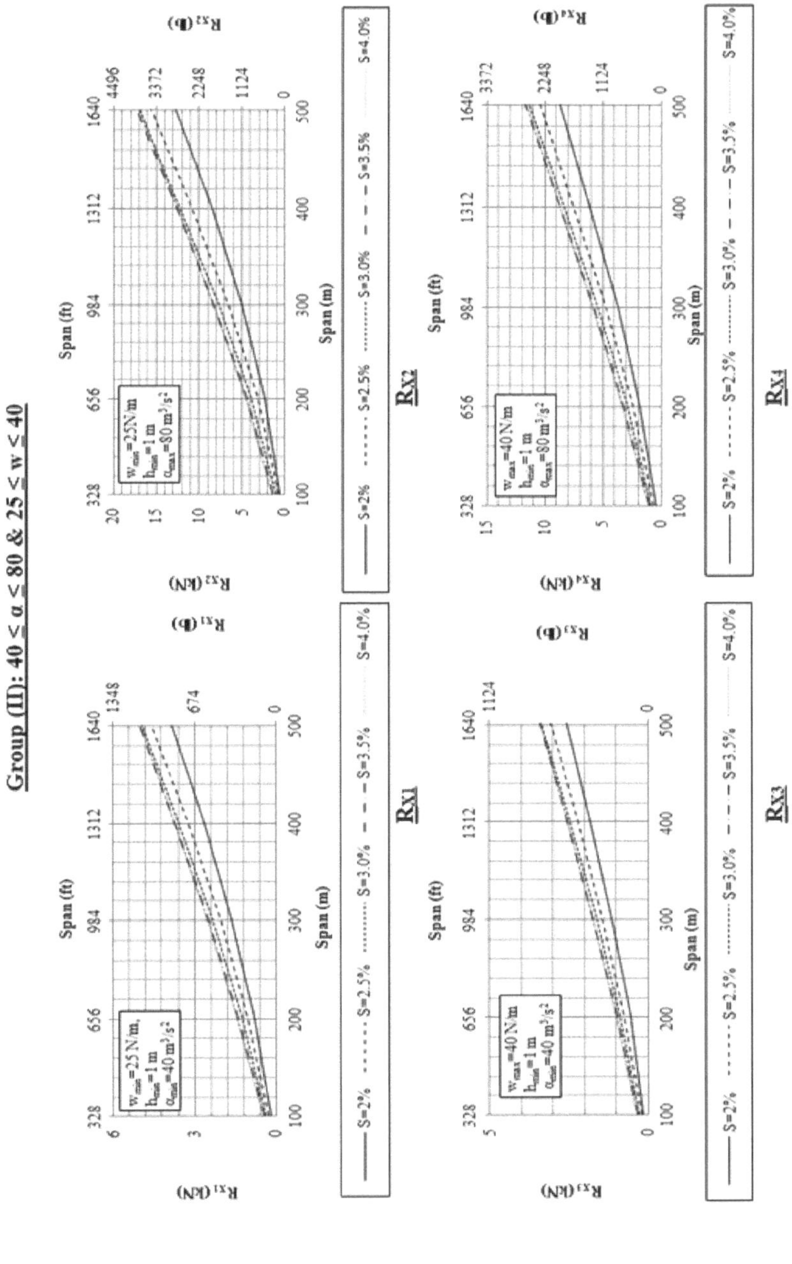

Figure K-8. Longitudinal force (R_X) charts: Group II.

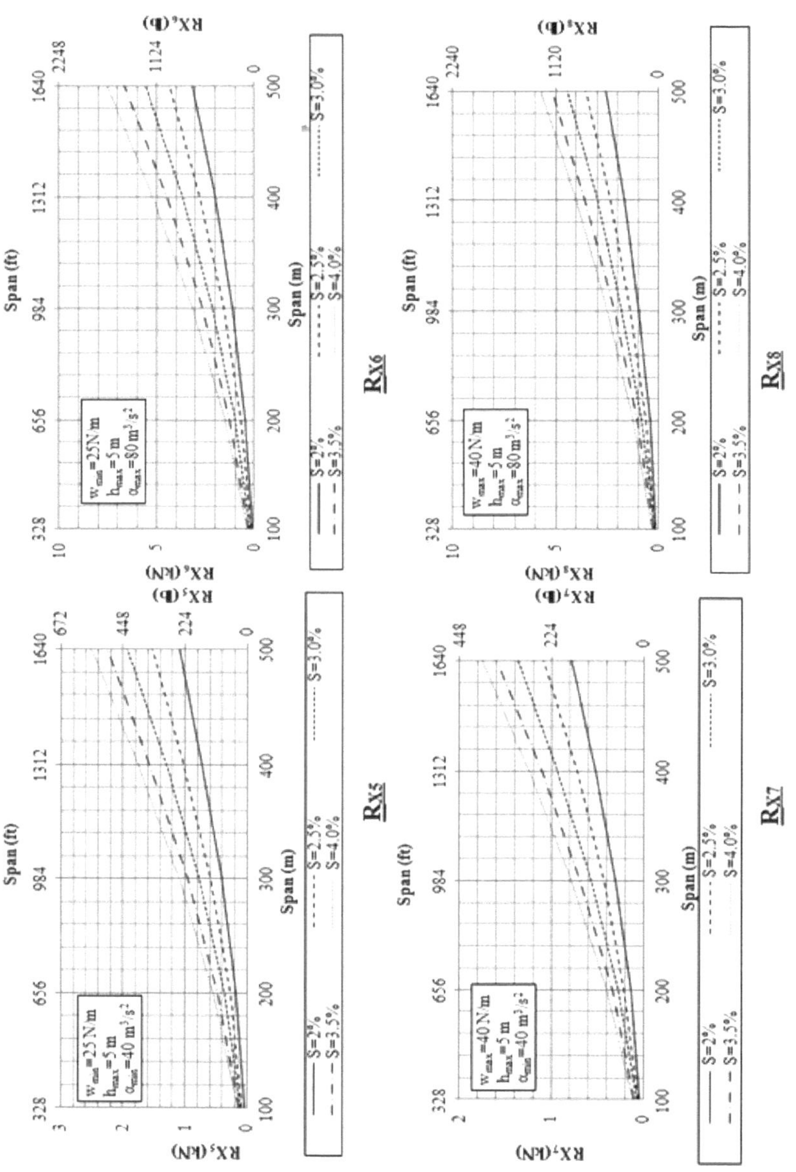

Figure K-8 (Continued). Longitudinal force (R_X) charts: Group II.

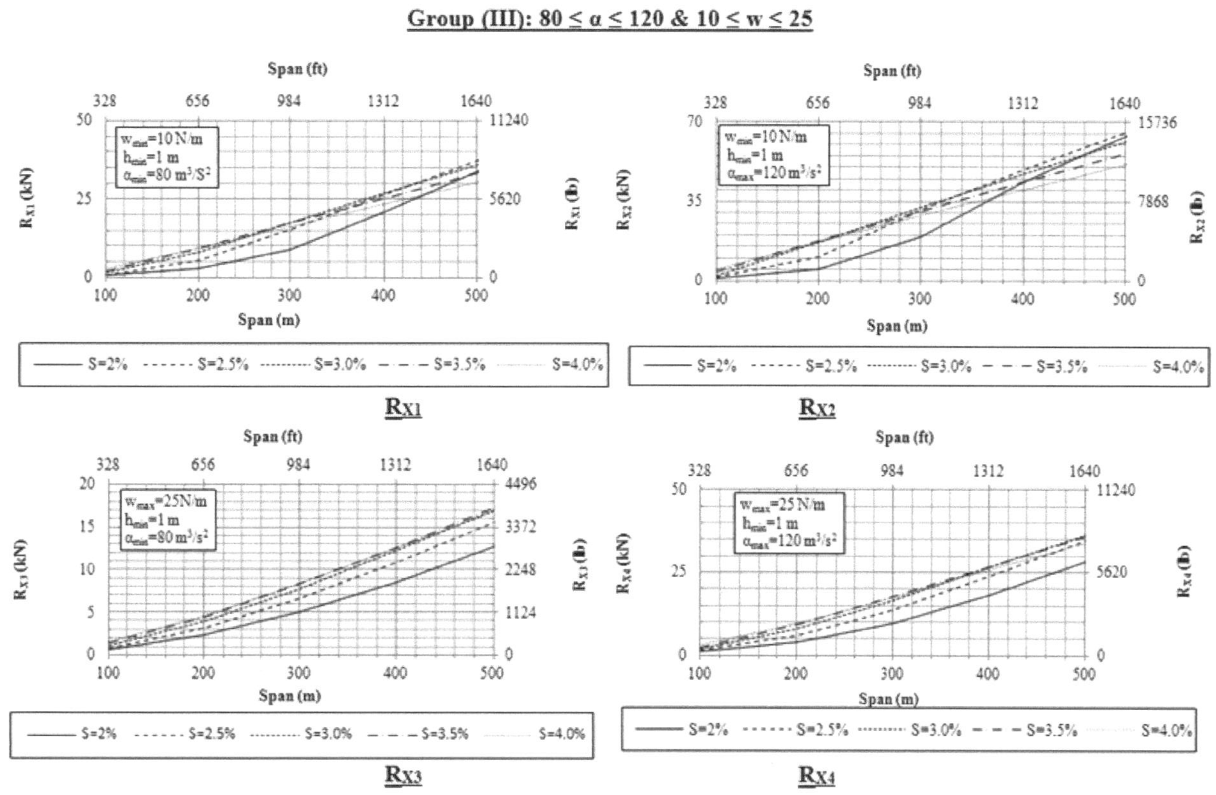

Figure K-9. Longitudinal force (R_X) charts: Group III.

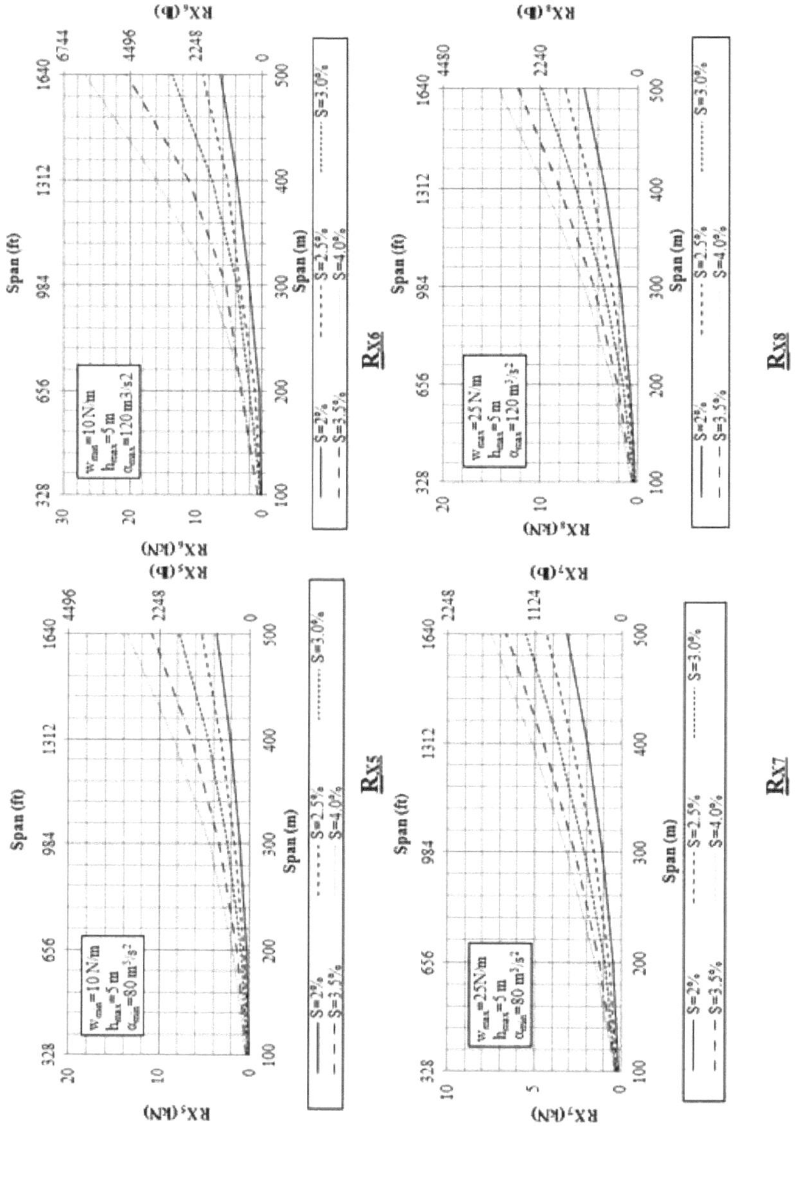

Figure K-9 (Continued). Longitudinal force (R_X) charts: Group III.

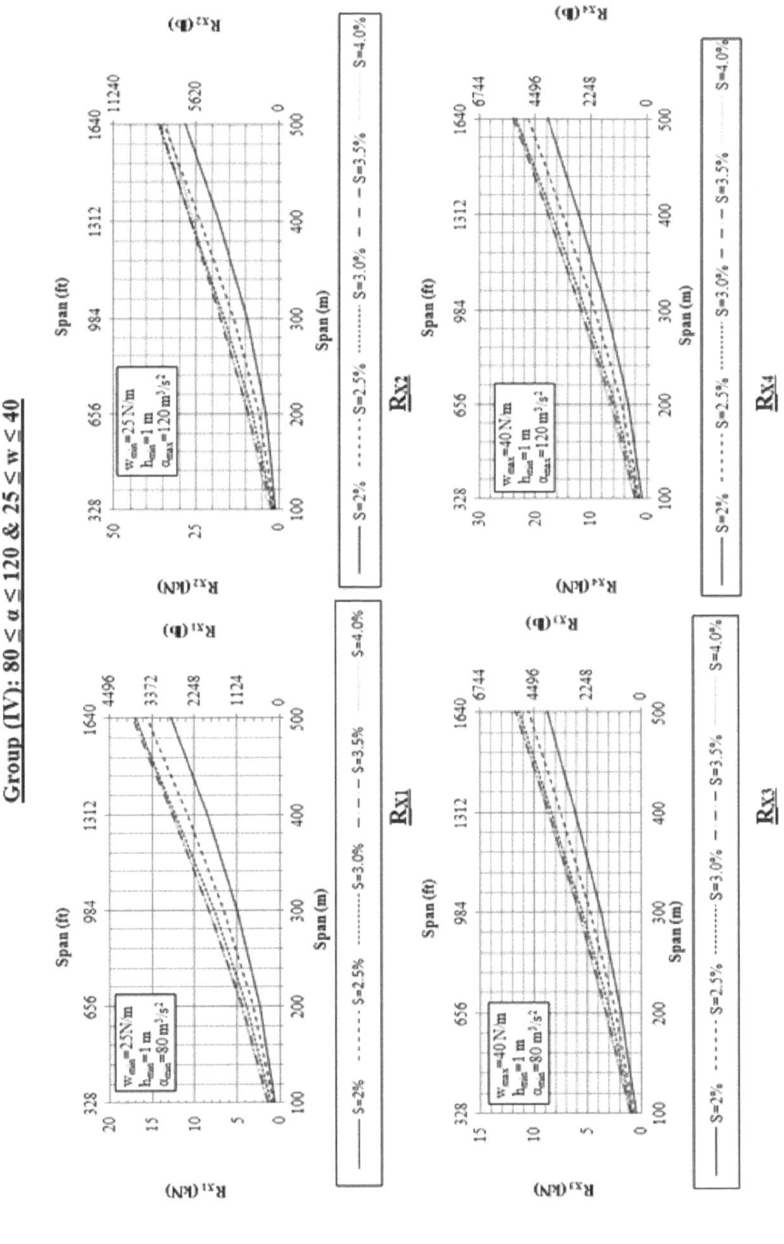

Figure K-10. Longitudinal force (R_X) charts: Group IV.

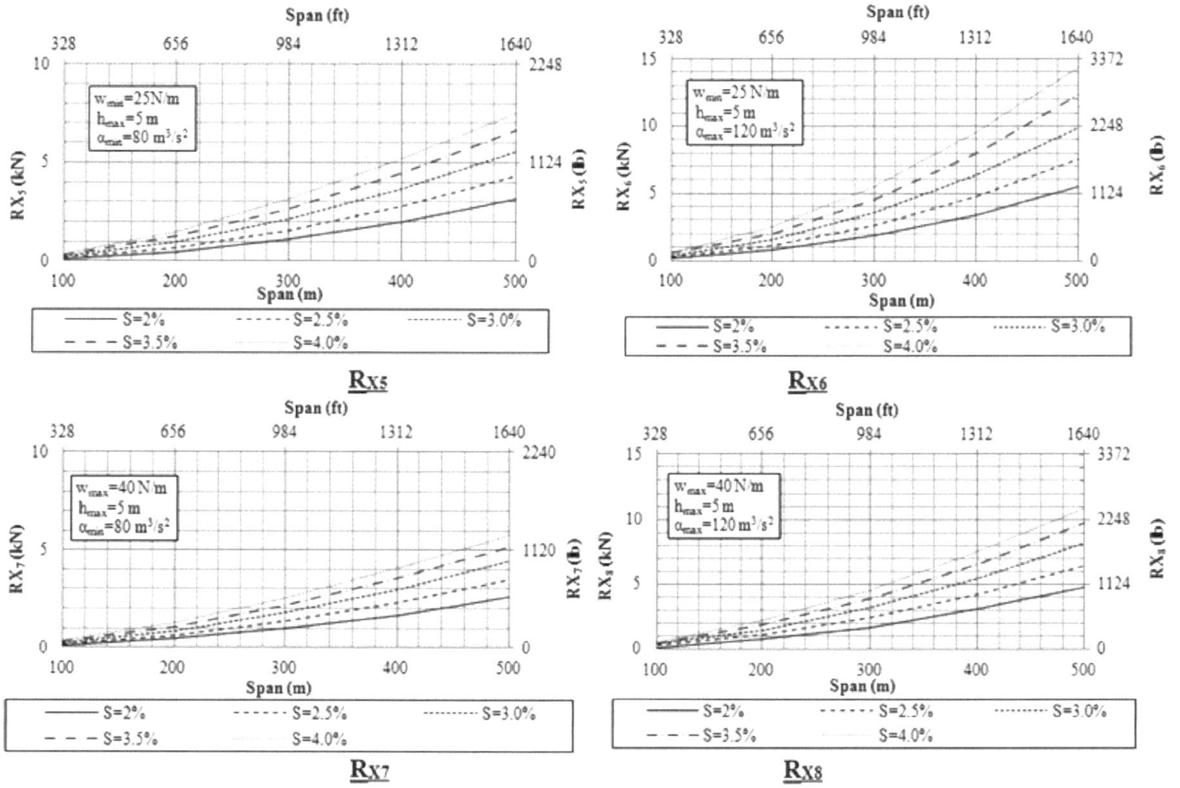

Figure K-10 (Continued). Longitudinal force (R_X) charts: Group IV.

Group (V): $120 \leq \alpha \leq 160$ & $10 \leq w \leq 25$

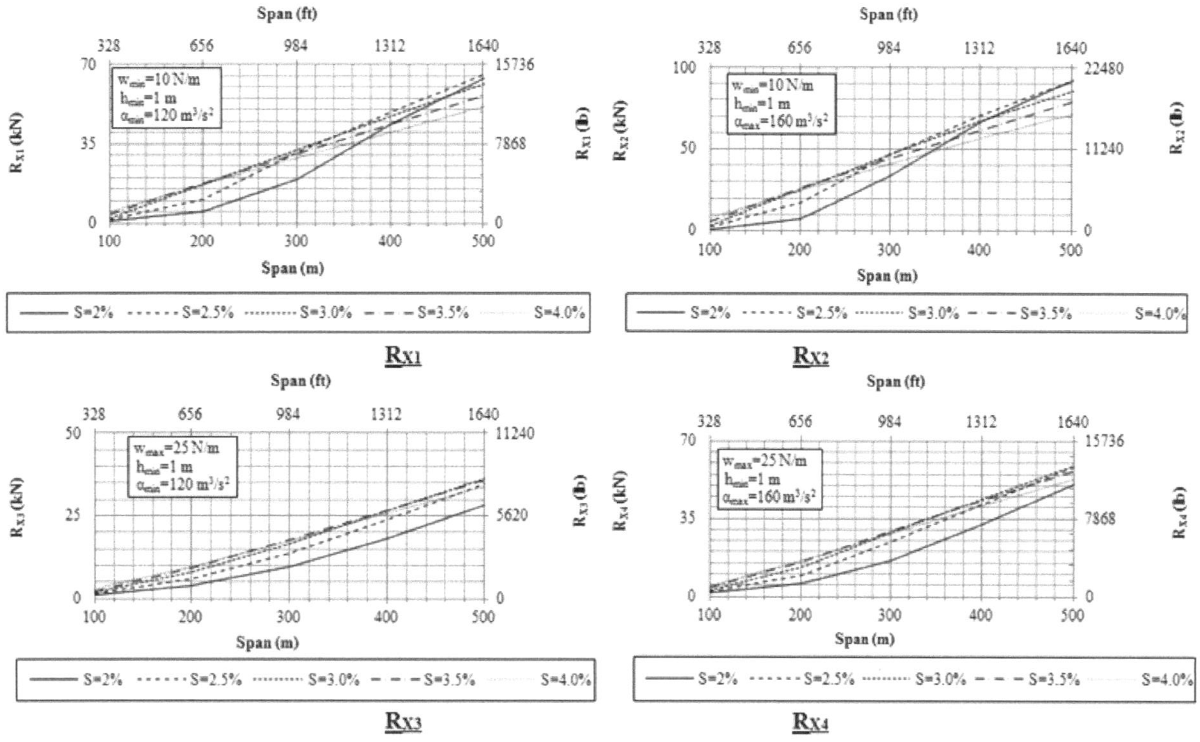

Figure K-11. Longitudinal force (R_X) charts: Group V.

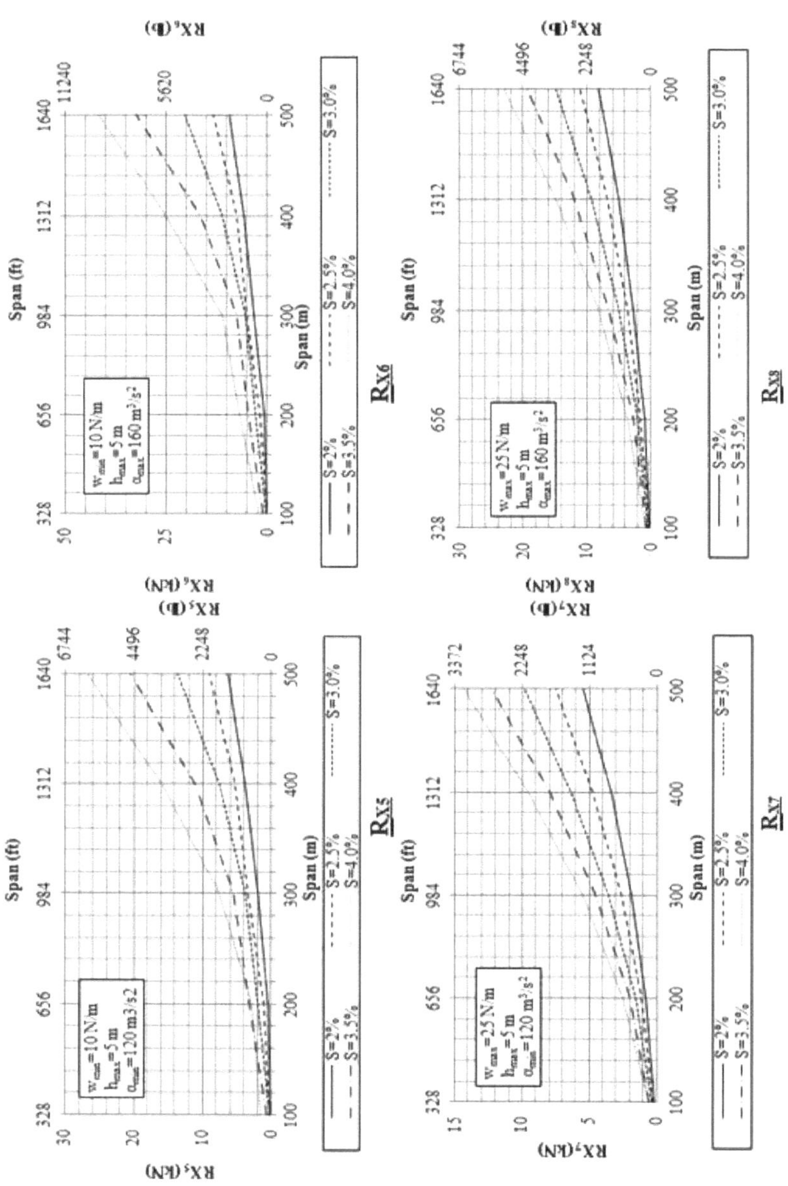

Figure K-11 (Continued). Longitudinal force (R_X) charts: Group V.

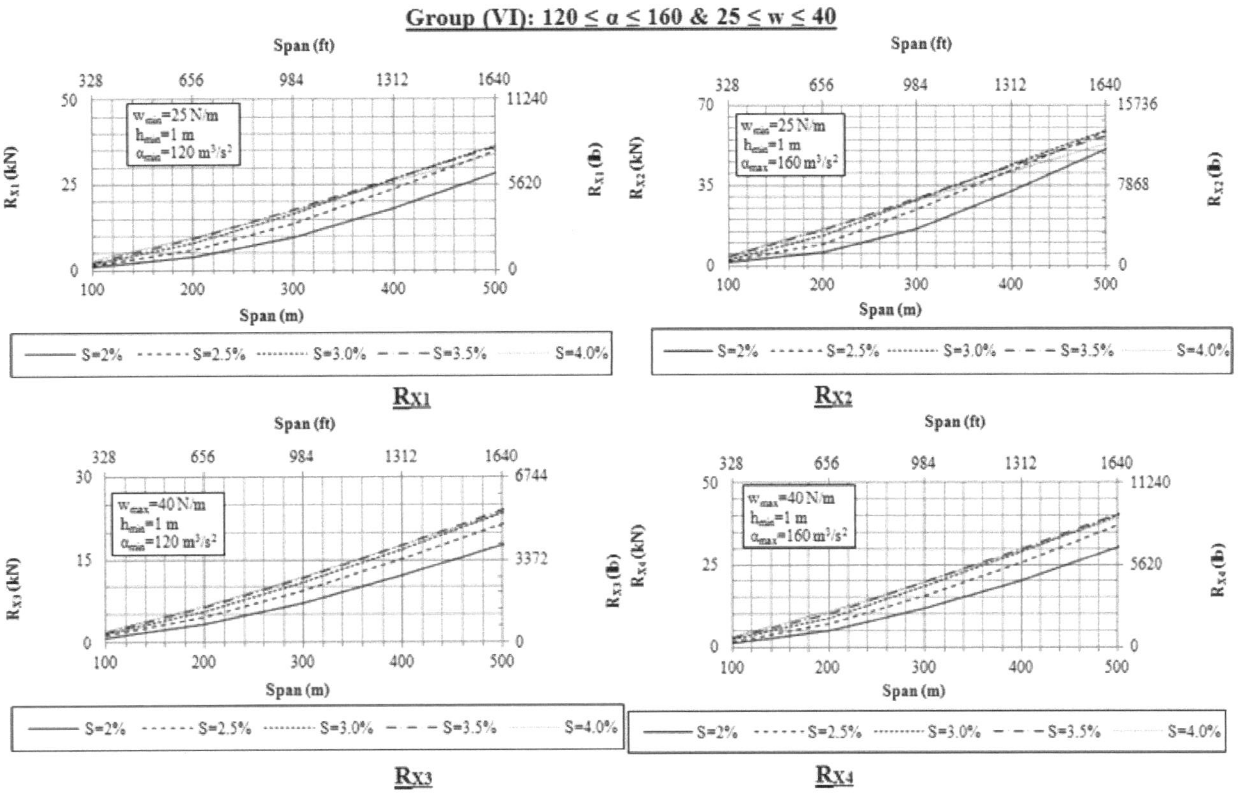

Figure K-12. Longitudinal force (R_X) charts: Group VI.

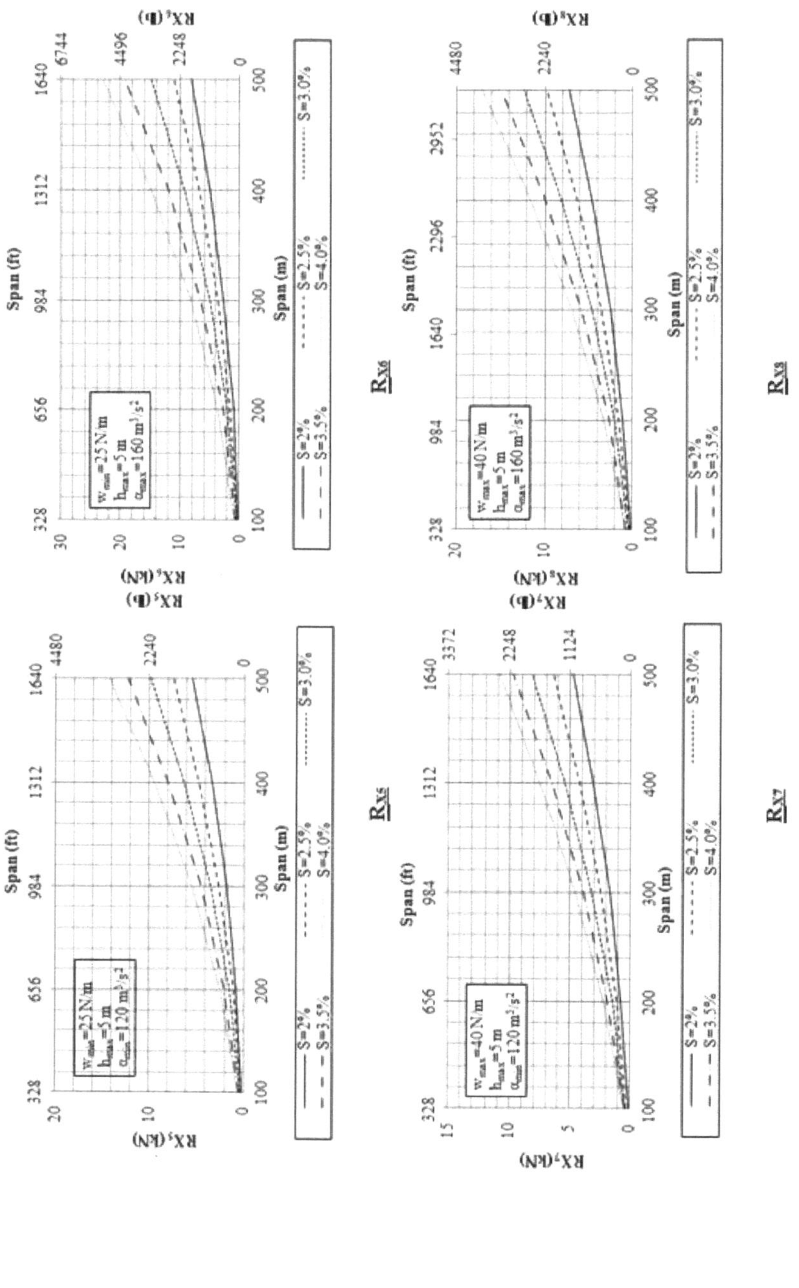

Figure K-12 (Continued). Longitudinal force (R_X) charts: Group VI.

Group (VII): 160 ≤ α ≤ 200 & 10 ≤ w ≤ 25

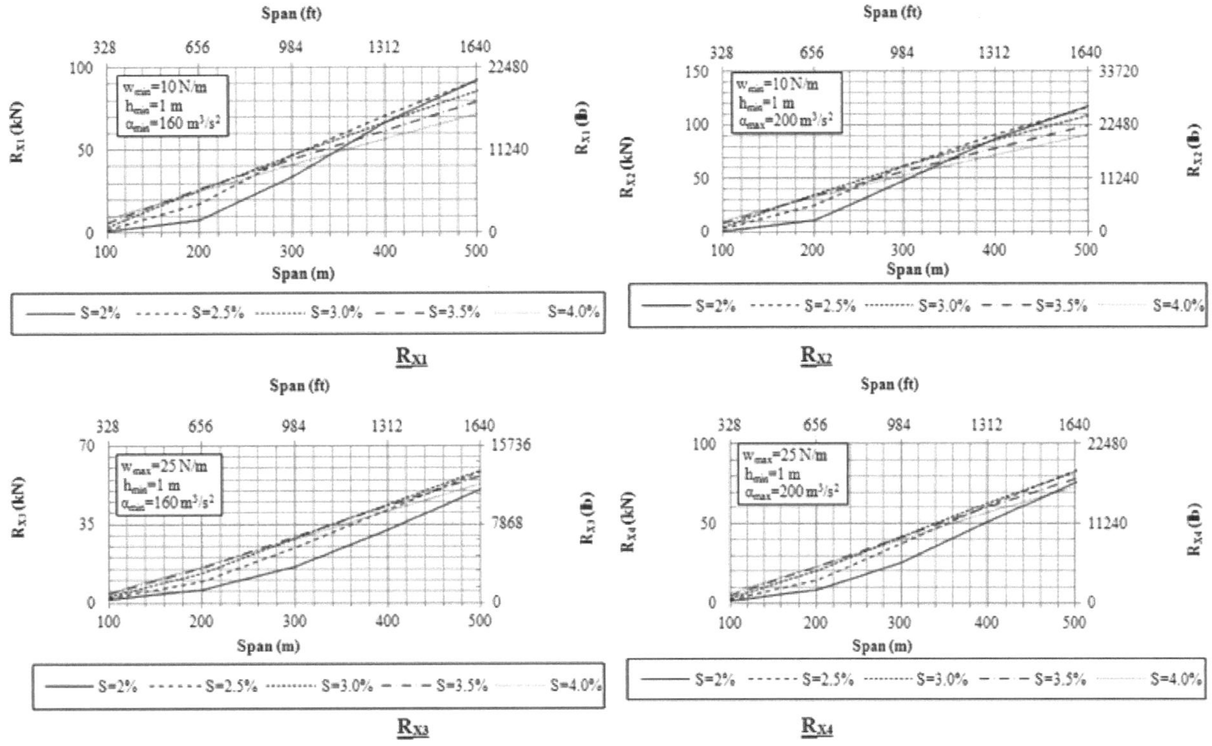

Figure K-13. Longitudinal force (R_X) charts: Group VII.

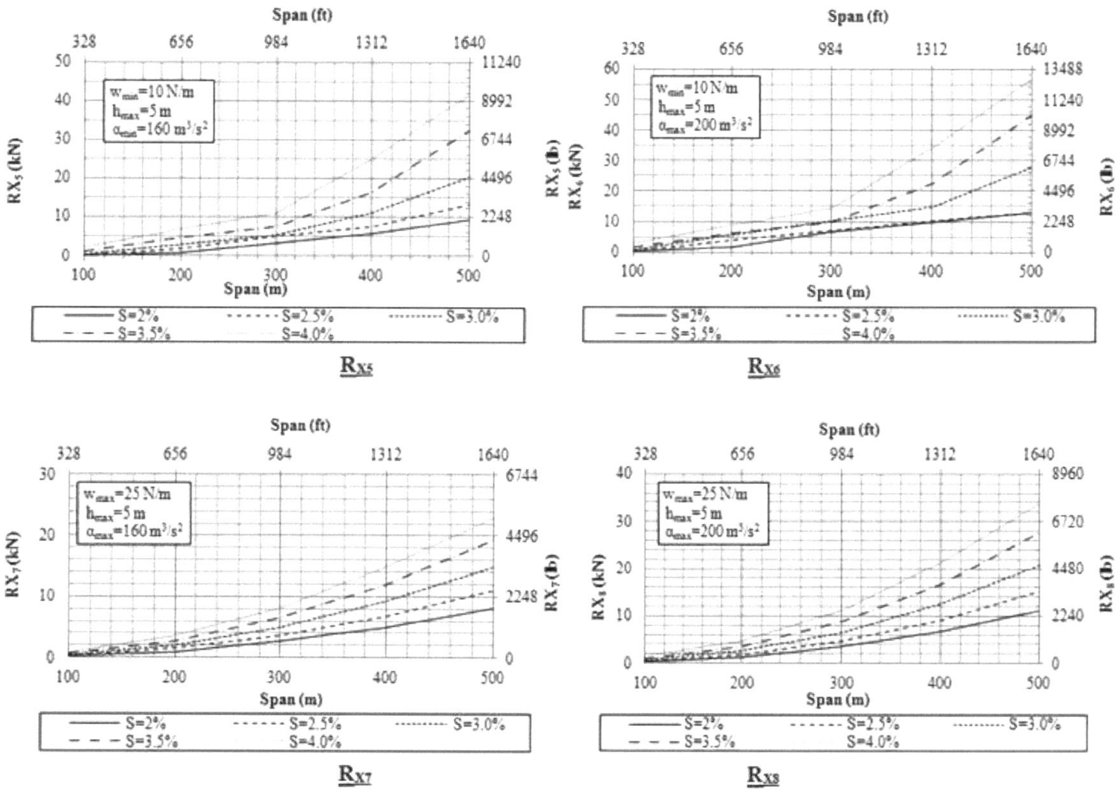

Figure K-13 (Continued). Longitudinal force (R_X) charts: Group VII.

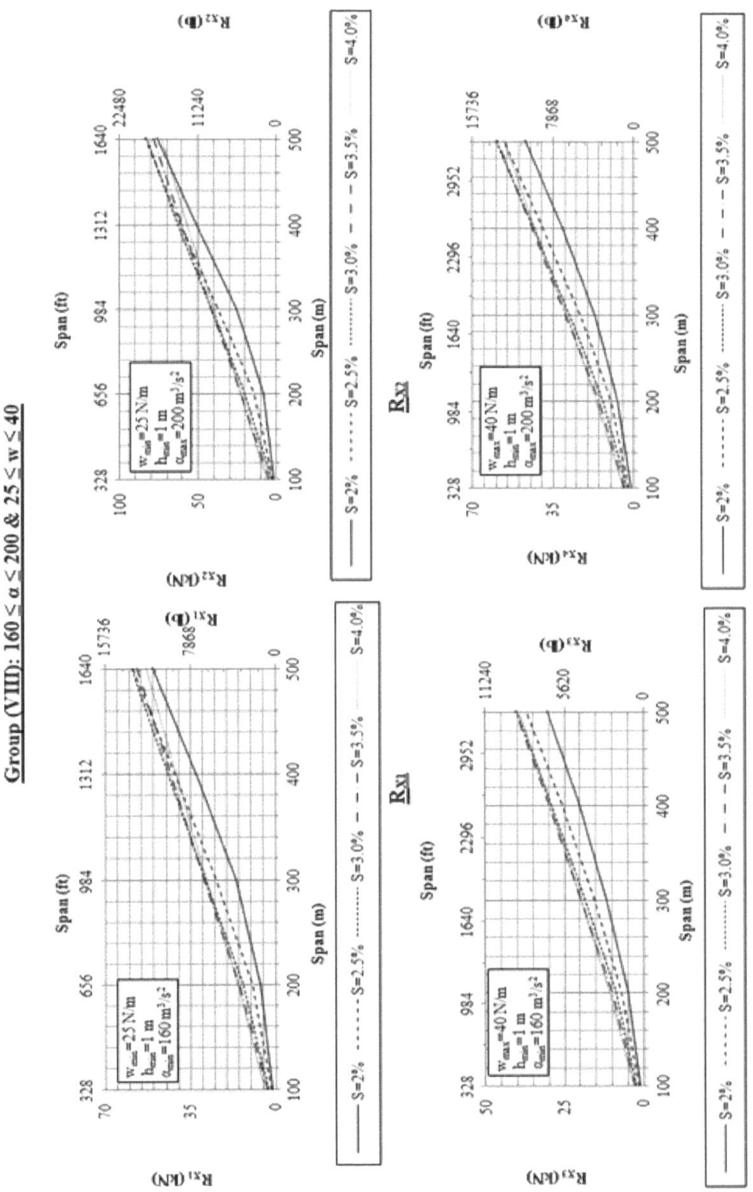

Figure K-14. Longitudinal force (R_X) charts: Group VIII.

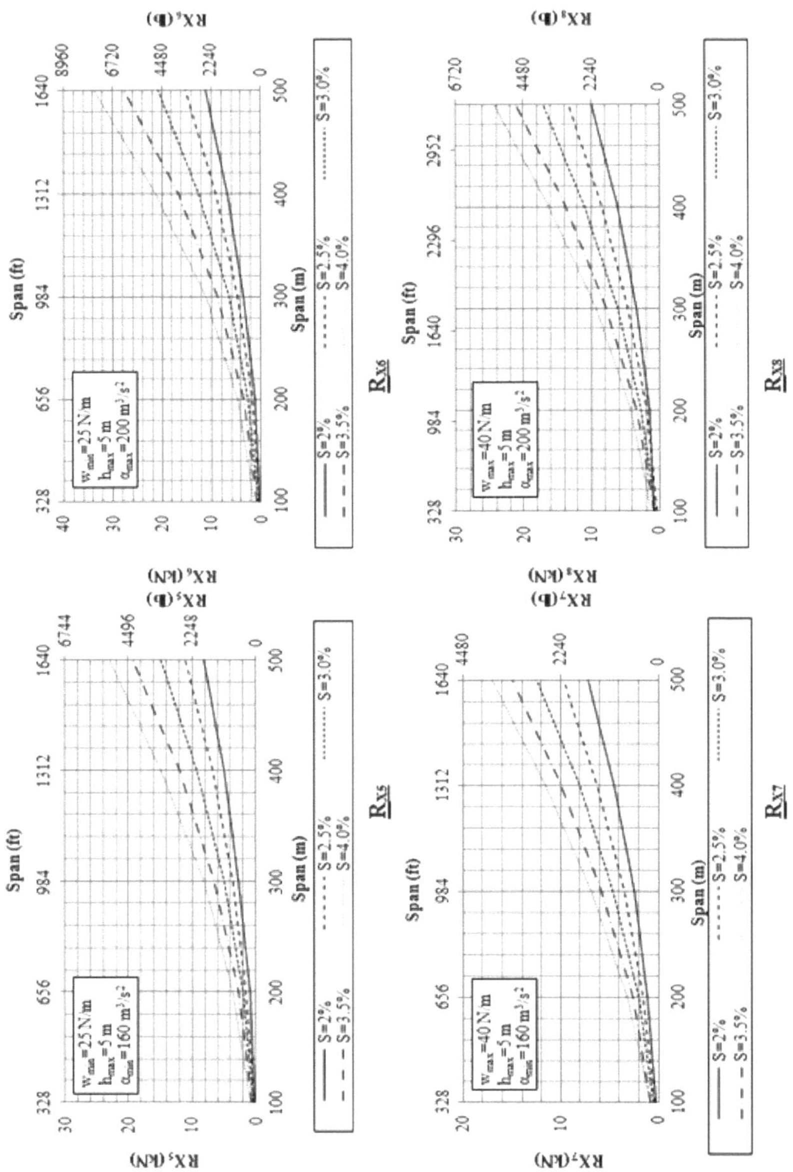

Figure K-14 (Continued). Longitudinal force (R_X) charts: Group VIII.

K.2.2.2 Shield Wire Longitudinal Force The main difference between a shield wire and a conductor is that a shield wire is attached to the structure directly or with a very short assembly. Separate design charts for the longitudinal unbalanced forces are developed for the shield wires (Elawady and El Damatty 2015). The value of the longitudinal forces are found to depend on the parameters *α*, *w*, *L*, and *S* described previously, in addition to the axial stiffness of the shield wire *EA*, where *E* is the modulus of elasticity and *A* is the cross-sectional area of the wire. The range of shield wire parameters considered are found in Table K-3.

Table K-3. Parameters Considered in the Shield Wire Design Charts

	Lower range		Upper range	
Parameter	SI units	US customary units	SI units	US customary units
α	30 m^3/sec^2	0.0933 $mile^3/hr^2$	100 m^3/sec^2	0.311 $mile^3/hr^2$
Cable weight w	3 N/m	0.205 lb/ft	15 N/m	1.028 lb/ft
Axial stiffness EA	0.9×10^7 N	2.02×10^6 lb	6×10^7 N	1.35×10^7 lb
Cable span L	100 m	328 ft	500 m	1640 ft
Sag ratio (sag/span ratio) S	2.5%	2.5%	4%	4%

The following steps can be conducted to obtain the longitudinal force associated with Load Case 3 for shield wires:

1. Calculate $\alpha = V_J^{2 \times} d$.
2. Based on the span value and the sag ratio, determine the longitudinal forces associated with the eight graphs of the design group (Figure K-15). Those are labeled
   R_{X1} = Longitudinal force corresponding to (w_{min}, EA_{min}, α_{min})
   R_{X2} = Longitudinal force corresponding to (w_{min}, EA_{min}, α_{min})
   R_{X3} = Longitudinal force corresponding to (w_{min}, EA_{min}, α_{min})
   R_{X4} = Longitudinal force corresponding to (w_{min}, EA_{min}, α_{min})
   R_{X5} = Longitudinal force corresponding to (w_{min}, EA_{min}, α_{min})
   R_{X6} = Longitudinal force corresponding to (w_{min}, EA_{min}, α_{min})
   R_{X7} = Longitudinal force corresponding to (w_{min}, EA_{min}, α_{min})
   R_{X8} = Longitudinal force corresponding to (w_{min}, EA_{min}, α_{min})
   Note: $EA_{max} = 6 \times 10^7$ N, $EA_{min} = 0.9 \times 10^7$ N, w_{max} = 15 N/m, w_{min} = 3 N/m, α_{max} = 100 m^3/sec^2, and α_{min} = 30 m^3/sec^2.

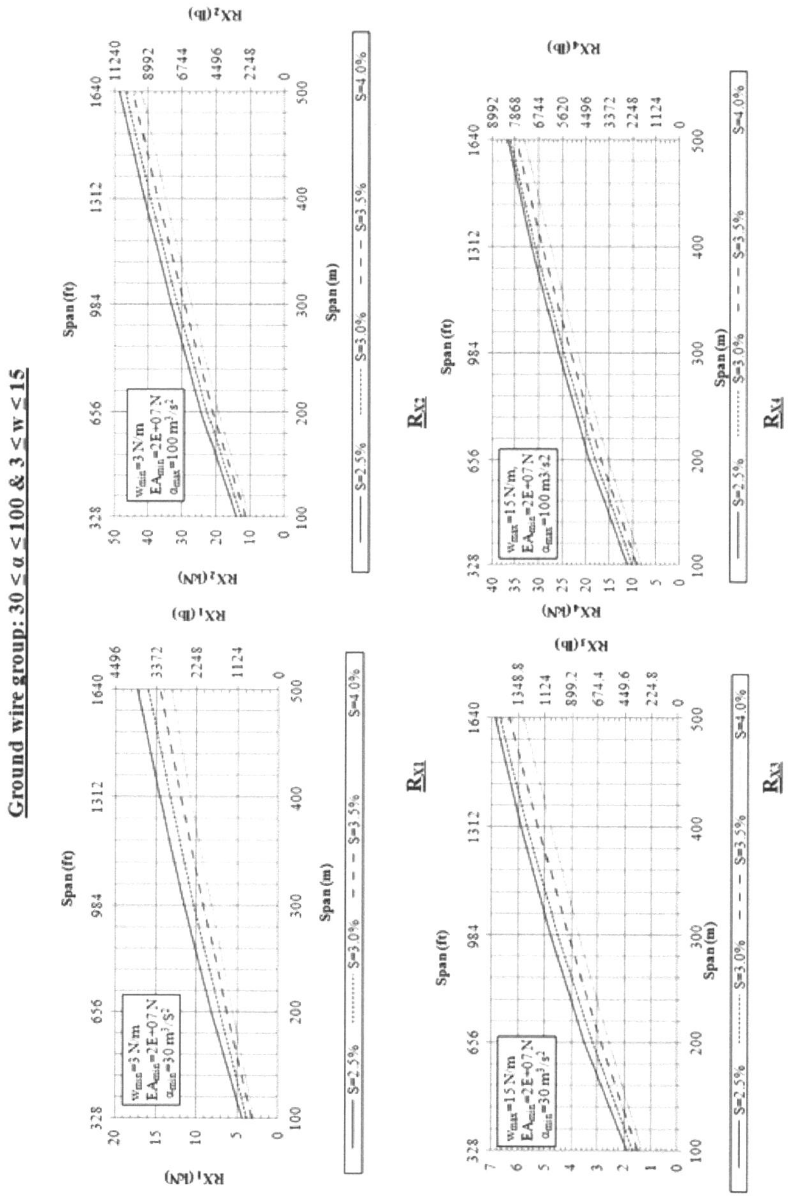

Figure K-15. Longitudinal force (R_X) charts: Shield wire group.

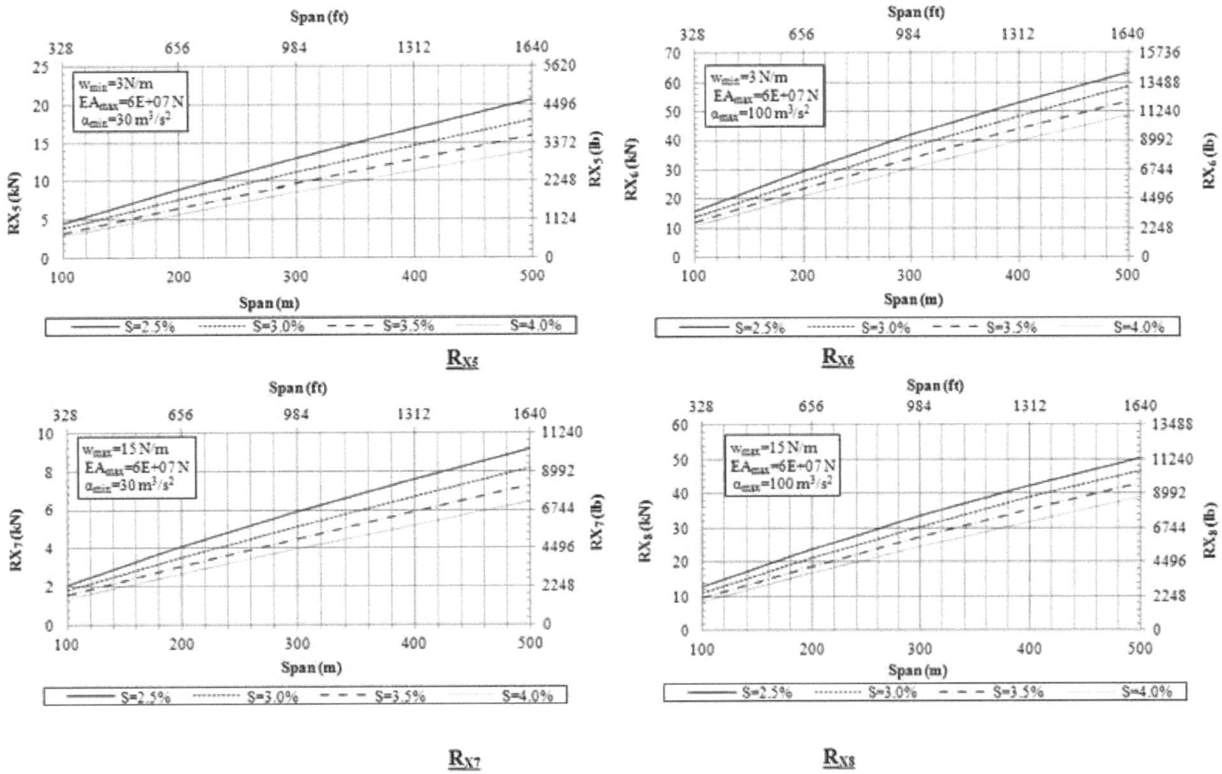

Figure K-15 (Continued). Longitudinal force (R_X) charts: Shield wire group.

3. Based on the shield wire values of α, EA, and w, and the eight evaluated longitudinal forces, linear interpolation can be performed using this set of equations:

$$R_{X(5-6)} = R_{X5} + (R_{X6} - R_{X5}) \times \frac{(\alpha - \alpha_{min})}{(\alpha_{max} - \alpha_{min})} \quad \text{(K-8)}$$

$$R_{X(7-8)} = R_{X7} + (R_{X8} - R_{X7}) \times \frac{(\alpha - \alpha_{min})}{(\alpha_{max} - \alpha_{min})} \quad \text{(K-9)}$$

$$R_{X(EA_{max})} = R_{X(7-8)} + (R_{X(5-6)} - R_{X(7-8)}) \times \frac{(w_{max} - w)}{(w_{max} - w_{min})} \quad \text{(K-10)}$$

$$R_X = R_{X(EA_{min})} + \frac{(R_{X(EA_{max})} - R_{X(EA_{min})})}{(EA_{max} - EA_{min})} \times (EA - EA_{min}) \quad \text{(K-11)}$$

$$R_{X(7-8)} = R_{X7} + (R_{X8} - R_{X7}) \times \frac{(\alpha - \alpha_{min})}{(\alpha_{max} - \alpha_{min})} \quad \text{(K-12)}$$

$$R_{X(EA_{max})} = R_{X(7-8)} + (R_{X(5-6)} - R_{X(7-8)}) \times \frac{(w_{max} - w)}{(w_{max} - w_{min})} \quad \text{(K-13)}$$

$$R_X = R_{X(EA_{min})} + \frac{(R_{X(EA_{max})} - R_{X(EA_{min})})}{(EA_{max} - EA_{min})} \times (EA - EA_{min}) \quad \text{(K-14)}$$

where R_x is the critical longitudinal unbalanced force resulting from this downburst load case and is to be applied at the shield wire to structure connection.

K.2.3 Example: Downburst Load Case 3

This example shows calculations of the conductor longitudinal unbalanced force under oblique downburst loading. The loads are based on the structure shown in Figure 5-1, the data are provided in Tables 5-1 to 5-3, and the design data are as follows:

Design Data

- Wind span = 1,500 ft = 457.2 m;
- Length of insulator assembly = 6 ft = 1.83 m;
- Conductor self-weight = 1.075 lb/ft = 15.7 N/m;
- Conductor projected diameter = 1.165 in. = 0.03 m;
- Line sag = 36 ft = 11 m (~2.5% span); and
- Assumed downburst jet velocity of 112 mph = 50 m/s.

Based on the aforementioned data, the following calculations are performed:

$$\alpha = V_J^2 \times d = 50^2 \times 0.03 = 75\ \text{m}^3/\text{s}^2(0.2332\ \text{mi}^2/\text{h}^2)$$

$$\Rightarrow (40(0.1244\ \text{mi}^3/\text{h}^2)) < \alpha < (80(0.2487\ \text{mi}^3/\text{hr}^2))$$

and

$$w = 15.7\ \text{N/m}\ (1.08\ \text{lb/ft})$$

$$\Rightarrow (10(0.685\ \text{lb/ft})) < w < (25(1.713\ \text{lb/ft}))$$

$$\Rightarrow \text{Group I, Figure K-7.}$$

Based on the charts given in Figure K-7 for Group I, the following values are extracted:

$R_{X1} = 1.8$ kips
$R_{X2} = 7.3$ kips
$R_{X3} = 0.85$ kips
$R_{X4} = 3.0$ kips
$R_{X5} = 0.4$ kips
$R_{X6} = 1.0$ kips
$R_{X7} = 0.3$ kips
$R_{X8} = 0.78$ kips

The following calculations are then performed:

$$R_{X(1-2)} = R_{X1} + (R_{X2} - R_{X1}) \times \frac{(\alpha - \alpha_{\min})}{(\alpha_{\max} - \alpha_{\min})}$$

$$\Rightarrow R_{X(1-2)} = 1.8 + (7.2 - 1.8) \times \frac{(0.2332 - 0.1244)}{(0.2487 - 0.1244)} = 6.52 \text{ kips}$$

$$R_{X(3-4)} = R_{X3} + (R_{X4} - R_{X3}) \times \frac{(\alpha - \alpha_{\min})}{(\alpha_{\max} - \alpha_{\min})}$$

$$\Rightarrow R_{X(3-4)} = 0.85 + (3.0 - 0.85) \times \frac{(0.2332 - 0.1244)}{(0.2487 - 0.1244)} = 2.68 \text{ kips}$$

$$R_{X(h\min)} = R_{X(3-4)} + (R_{X(1-2)} - R_{X(3-4)}) \times \frac{(w_{\max} - w)}{(w_{\max} - w_{\min})}$$

$$\Rightarrow R_{X(h\min)} = 2.68 + (6.52 - 2.68) \times \frac{(1.713 - 1.08)}{(1.713 - 0.685)} = 5.061 \text{ kips}$$

$$R_{X(5-6)} = R_{X5} + (R_{X6} - R_{X5}) \times \frac{(\alpha - \alpha_{\min})}{(\alpha_{\max} - \alpha_{\min})}$$

$$\Rightarrow R_{X(5-6)} = 0.4 + (1.0 - 0.4) \times \frac{(0.2332 - 0.1244)}{(0.2487 - 0.1244)} = 0.92 \text{ kips}$$

$$R_{X(7-8)} = R_{X7} + (R_{X8} - R_{X7}) \times \frac{(\alpha - \alpha_{min})}{(\alpha_{max} - \alpha_{min})}$$

$$\Rightarrow R_{X(7-8)} = 0.3 + (0.78 - 0.3) \times \frac{(0.2332 - 0.1244)}{(0.2487 - 0.1244)} = 0.73 \text{ kips}$$

$$R_{X(h\max)} = R_{X(7-8)} + (R_{X(5-6)} - R_{X(7-8)}) \times \frac{(w_{max} - w)}{(w_{max} - w_{min})}$$

$$\Rightarrow R_{X(h\max)} = 0.73 + (0.92 - 0.73) \times \frac{(1.713 - 1.08)}{(1.713 - 0.685)} = 0.84 \text{ kips}$$

Finally, the longitudinal unbalanced force is evaluated as follows:

$$R_X = R_{X(h\max)} + \frac{(R_{X(h\min)} - R_{X(h\max)})}{(h_{max} - h_{min})} \times (h_{max} - h)$$

$$\Rightarrow R_X = 0.84 + \frac{(5.061 - 0.84)}{(16.4 - 3.28)} \times (16.4 - 6) = 4.18 \text{ kips}$$

K.3 TORNADOES

K.3.1 F2 Tornado Critical Load Cases

Two methods of analysis to account for the critical effect of tornadoes on transmission line structures are provided in this appendix. It should be noted that the simplified analysis is applicable only for cantilever-type structures (self-supported), while the detailed analysis can be applied to any transmission line system.

K.3.1.1 Simplified Analysis For cantilever-type structures (self-supported), the critical effect of tornadoes is simulated by applying a uniform velocity along the height of the structure with a magnitude of 161 mph (72 m/sec) accompanied with a uniform velocity of 161 mph (72 m/sec) applied on the wires. This uniform velocity field is applied from all potentially sensitive directions. The effect of the yaw angle Ψ should follow the procedure described in Section 2.1.6.2. The forces acting on the wires due to this uniform velocity can be reduced using the span reduction factor (SRF) calculated using the Equation (K-15) (Behncke and Ho 2009):

$$SRF = W_G\,(1 - 0.25W_G/L)/L \quad \text{(K-15)}$$

where W_G is the tornado gust width with a recommended value of 490 ft (150 m) for F2 tornadoes, and L is the average span length of the transmission line.

K.3.1.2 Detailed Analysis Two simplified load cases that simulate the critical effect of F2 tornadoes on transmission structures, which can be applied to both self-supported and guyed structures, are provided as follows (El Damatty et al. 2015). Each load case has two vertical velocity profiles along the structure height combined with a uniformly distributed transverse wind velocity profile on the conductors as shown.

K.3.1.2.1 Load Case 1 This load case is based on the following three wind profiles:

1. Vertical velocity profile A acting on one face of the structure shown in Figure K-16.
2. Vertical velocity profile B acting on the perpendicular face of the structure shown in Figure K-17.
3. Profile C of uniform velocity distribution on the conductors with a gust value of 161 mph (72 m/sec). The forces acting on the wires due to this uniform velocity can be reduced using the span reduction factor (SRF) calculated using Equation (K-15).

It is not required to consider all potential wind directions in this load case. Only the wind loading profile combinations shown in Figure K-18 are to be considered in the application of this load case.

K.3.1.2.2 Load Case 2 This case corresponds to a tornado center located on an oblique angle relative to the line. This load case is based on the following three wind profiles:

1. Vertical velocity profile D acting on one face of the structure shown in Figure K-19.
2. Vertical velocity profile E acting on the perpendicular face of the structure shown in Figure K-20.
3. Profile F of uniform velocity distribution on the conductors with an amplitude 161 mph (72 m/sec). The forces acting on the wires due to this uniform velocity can be reduced using the span reduction factor (SRF) calculated using Equation (K-15).

It is not required to consider all potential wind directions in this load case. Only the wind loading profile combinations shown in Figure K-21 are to be considered in the application of this load case.

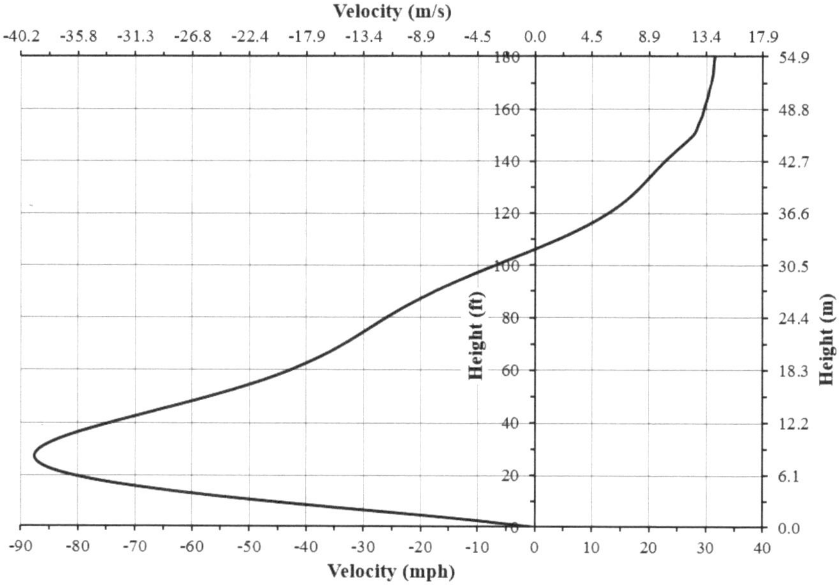

Figure K-16. Vertical Velocity Profile A.

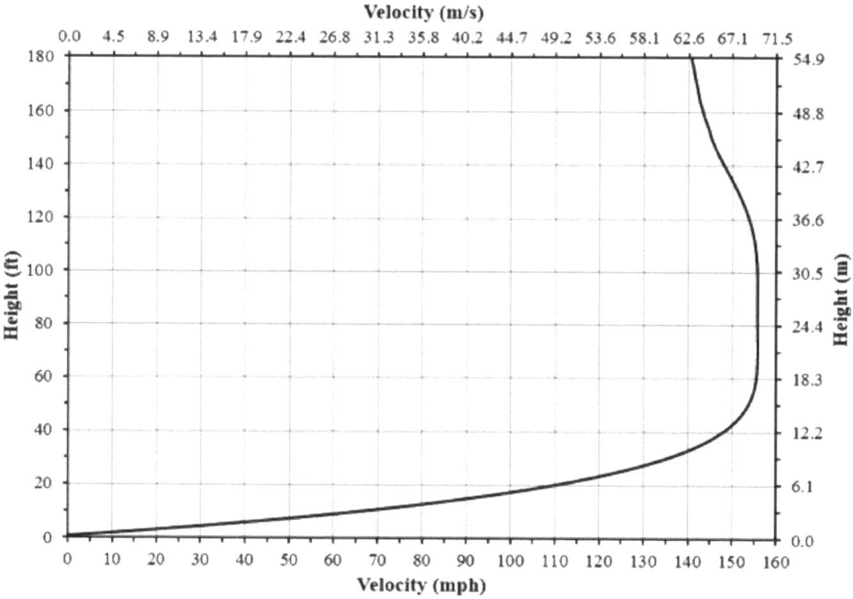

Figure K-17. Vertical Velocity Profile B.

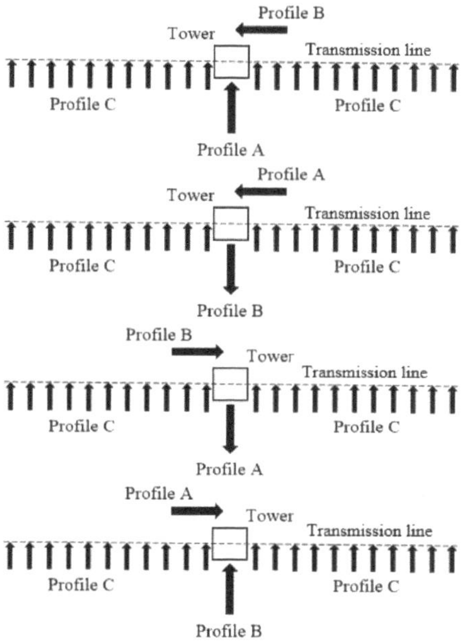

Figure K-18. Possible combinations of the vertical wind profiles A and B for Load Case 1.

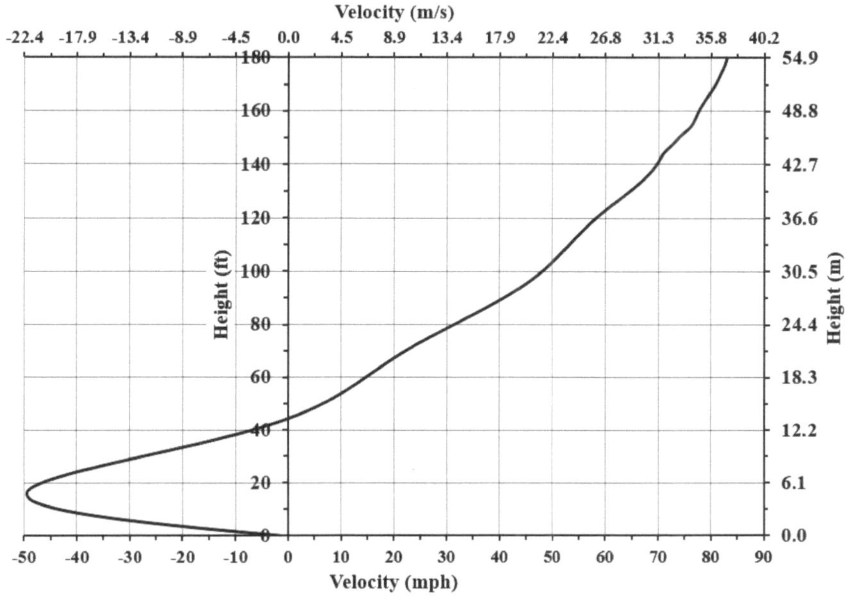

Figure K-19. Vertical Velocity Profile D.

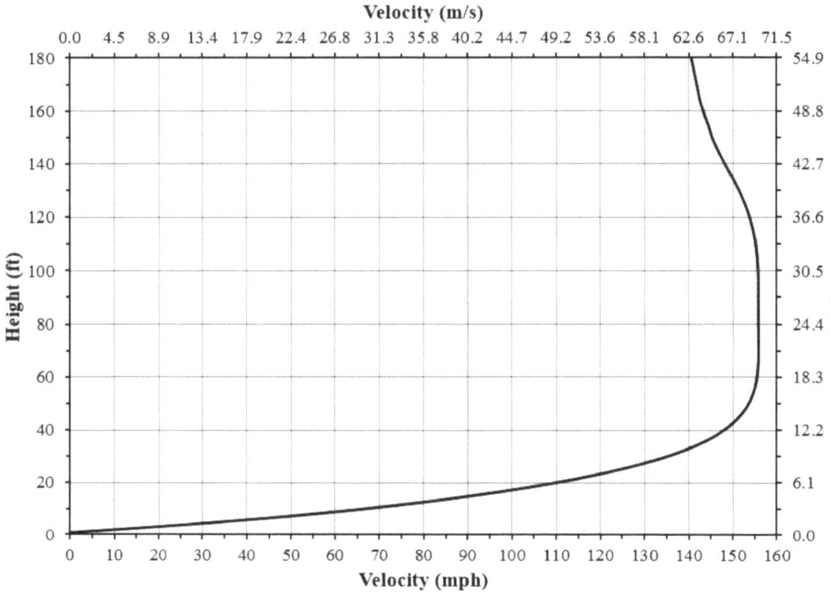

Figure K-20. Vertical Velocity Profile E.

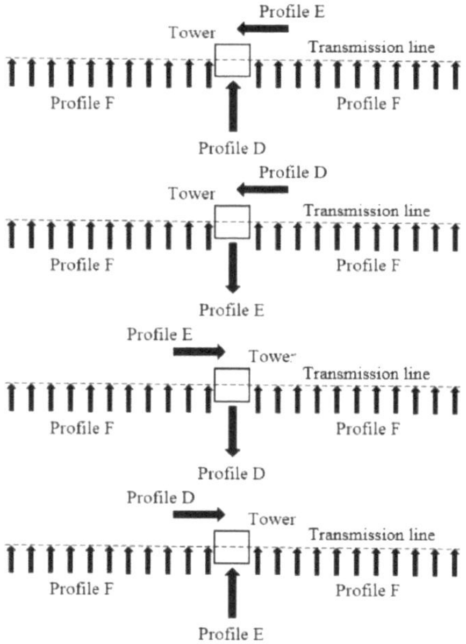

Figure K-21. Possible combinations of the vertical wind profiles D and E.

For a structure, the following combinations are to be considered:

Tornado wind speeds are gust wind speeds; therefore, the gust response factor, G, and velocity pressure exposure coefficient, K_z, should be considered equal to 1.0 for calculating the wind force.

Transmission structures may be susceptible to wind-borne debris initiated by HIW. These load cases do not account for debris impact loads.

K.4 HIW NOMENCLATURE

K.4.1 Downburst Section

d Single wire projected diameter perpendicular to the radial direction of the downburst velocity;

D_J Downburst jet diameter;

h Insulator length;

R Distance from the center of the downburst to the center of the structure of interest;

R_X Downburst critical longitudinal unbalanced force based on the proposed design charts corresponding to the actual insulator length;

$R_{X\,(hmax)}$ Wire longitudinal force corresponding to the selected design group maximum insulator length, the line's actual α, and the line's actual w;

$R_{X\,(hmin)}$ Wire longitudinal force corresponding to the selected design group minimum insulator length, the line's actual α, and the line's actual w;

R_{X1} Wire longitudinal force under Downburst Critical Case 3 corresponding to specific design group (w_{min}, h_{min}, α_{min});

R_{X2} Wire longitudinal force under Downburst Critical Case 3 corresponding to specific design group (w_{max}, h_{max}, α_{max});

R_{X3} Wire longitudinal force under Downburst Critical Case 3 corresponding to specific design group (w_{min}, h_{min}, α_{min});

R_{X4} Wire longitudinal force under Downburst Critical Case 3 corresponding to specific design group (w_{max}, h_{max}, α_{max});

R_{X5} Wire longitudinal force under Downburst Critical Case 3 corresponding to specific design group (w_{min}, h_{min}, α_{min});

R_{X6} Wire longitudinal force under Downburst Critical Case 3 corresponding to specific design group (w_{max}, h_{max}, α_{max});

R_{X7} Wire longitudinal force under Downburst Critical Case 3 corresponding to specific design group (w_{min}, h_{min}, α_{min});

R_{X8} Wire longitudinal force under Downburst Critical Case 3 corresponding to specific design group (w_{max}, h_{max}, α_{max});

$R_{X(1\text{-}2)}$ Wire longitudinal force corresponding to the selected design group minimum weight, minimum insulator length, and the line's actual α;
$R_{X(3\text{-}4)}$ Wire longitudinal force corresponding to the selected design group maximum weight, minimum insulator length, and the line's actual α;
$R_{X(5\text{-}6)}$ Wire longitudinal force corresponding to the selected design group minimum weight, maximum insulator length, and the line's actual α;
$R_{X(7\text{-}8)}$ Wire longitudinal force corresponding to the selected design group maximum weight, maximum insulator length, and the line's actual α;
R_Y Wire force in the transverse direction;
S Wire sag divided by line span;
V_{eqc} Equivalent uniform velocity distribution for wires;
V_{eqt} Equivalent uniform velocity distribution for structures;
V_J Downburst jet velocity;
V_{RD} Radial component of the downburst outflow velocity;
V_{VR} Vertical component of the downburst outflow velocity;
w Single wire weight;
X-axis Axis that passes along the line direction (longitudinal direction);
Y-axis Axis perpendicular to the line direction (transverse direction);
α Product of V_J^2 and d; and
Ψ Angle between the vertical plane of the transverse direction and the vertical plane connecting the center of the downburst and the center of the structure of interest.

K.4.2 Tornado Section

L Average span length of the transmission line;
r Radial distance from tornado center; and
W_G Tornado gust width.

APPENDIX L
WEATHER-RELATED LOADS FOR ADDITIONAL MRIS

This Manual of Practice recommends that transmission lines be designed for weather-related loads (i.e., wind and ice) corresponding to a basic or reference mean recurrence interval (MRI) of 100 years. It is acknowledged that it may be of interest to design a transmission line to a level of reliability lower or higher than the reference MRI. This appendix provides the design data for selected MRIs and equations for loading to be consistent with Chapter 2.

Maps of basic wind speeds were developed for 50-year and 300-year MRIs (ASCE 2017) based on a methodology consistent with that discussed in Chapter 2 (Pintar et al. 2015). Similar maps of radial ice thickness from freezing rain with concurrent gust wind speeds were developed for 50-year and 300-year MRI, also based on a methodology consistent with that discussed in Chapter 2. It should be pointed out that all the notes, methods of application, and limitations apply to these maps as to those presented in Chapter 2, and users should familiarize themselves with these aspects.

L.1 50-YEAR MRI

The basic wind speed map for a 50-year MRI is shown in Figure L-1. Values are 3-second gust wind speeds in miles per hour (m/s also shown) at 33 ft (10 m) above ground in terrain with Exposure Category C. This map of wind speeds is associated with an annual exceedance probability of 0.02 (2%) and corresponds to approximately a 64% probability of exceedance in 50 years.

The entire state of Hawaii is defined as a Special Wind Region on the current wind speed maps. This is due to the extreme topographic conditions found throughout the Hawaiian Islands. Following a review of the K_{zt} maps for Hawaii, the following wind speeds are recommended for a MRI of 50 years:

1. Wind speed of 82 mph (37 m/s) for regions indicated as $K_{zt} \leq 1.5$
2. Wind speed of $\left(67\,\text{mph} \bullet \sqrt{K_{zt}}\right)$ or $\left(30\,\text{m/s} \bullet \sqrt{K_{zt}}\right)$ for regions indicated as $K_{zt} > 1.5$.

In areas where local historical icing data are not available, the glaze ice maps given in Figures L-2 through L-6 can be used with some limitations. These maps show 50-year MRI ice thicknesses due to freezing precipitation with concurrent 3-second gust wind speeds in miles per hour (m/s) at 33 ft (10 m) above ground for the continental United States and Alaska. The glaze ice thicknesses shown in these figures do not include in-cloud icing or sticky snow accretions, which are caused by meteorological conditions that may produce significantly different loading patterns (see Appendix H, Section H.5).

L.2 300-YEAR MRI

The basic wind speed map for a 300-year MRI is shown in Figure L-7. Values are 3-second gust wind speeds in miles per hour (m/s) at 33 ft (10 m) above ground in terrain with Exposure Category C. This map of wind speeds is associated with an annual exceedance probability of 0.00333 (0.333%) and corresponds to approximately a 15% probability of exceedance in 50 years.

The entire state of Hawaii is defined as a Special Wind Region on the current wind speed maps. This is due to the extreme topographic conditions found throughout the Hawaiian Islands. Following a review of the K_{zt} maps for Hawaii, the following wind speeds are recommended for a MRI of 300 years:

1. Wind speed of 141 mph (63 m/s) for regions indicated as $K_{zt} \leq 1.5$
2. Wind speed of $\left(115\,\text{mph} \bullet \sqrt{K_{zt}}\right)$ or $\left(51\,\text{m/s} \bullet \sqrt{K_{zt}}\right)$ for regions indicated as $K_{zt} > 1.5$.

In areas where local historical icing data are not available, the glaze ice maps given in Figures L-8 through L-12 can be used with some limitations. These maps show 300-year ice thicknesses due to freezing precipitation with concurrent 3-second gust wind speeds in miles per hour (m/s) at 33 ft (10 m) above ground for the continental United States and Alaska. The glaze ice thicknesses shown in these figures do not include in-cloud icing or sticky snow accretions, which are caused by meteorological conditions that may produce significantly different loading patterns (see Appendix H, Section H.5).

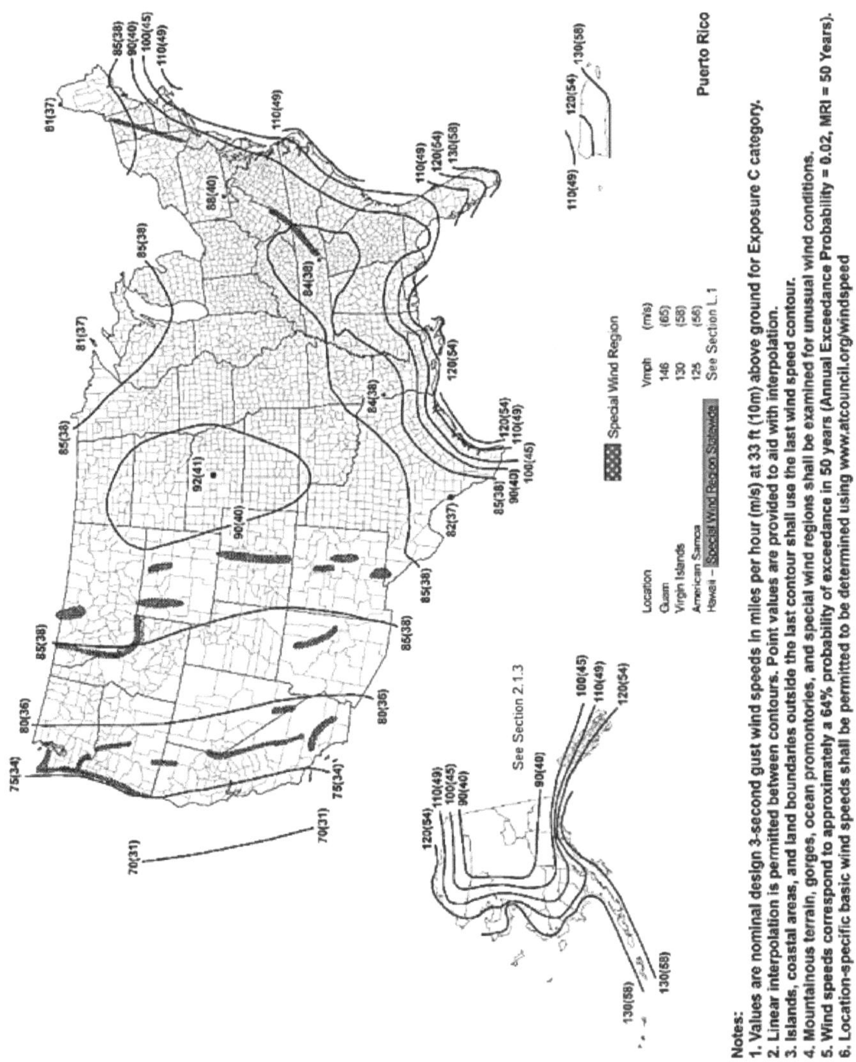

Figure L-1. 50-year MRI 3-second gust wind speed map [mph (m/s)] at 33 ft (10 m) aboveground in Exposure Category C.
Source: Minimum Design Loads and Associated Criteria for Buildings and Other Structures *(ASCE (2017).*

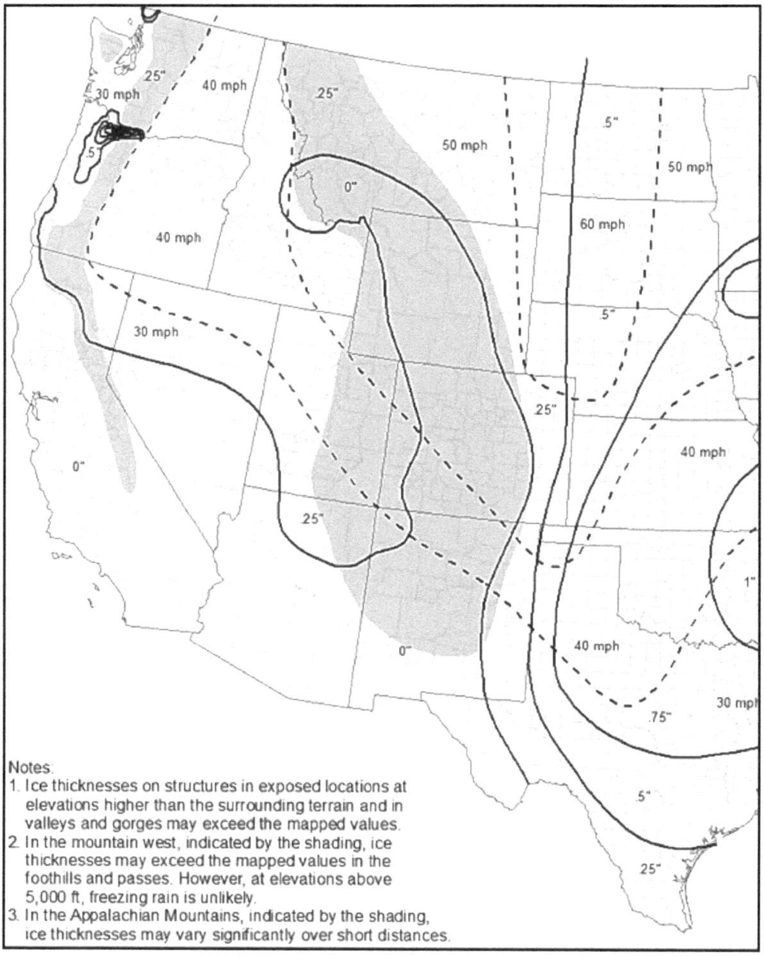

(a)

Figure L-2. 50-year MRI radial ice thickness (in.) from freezing rain with concurrent gust wind speeds (mph) at 33 ft (10 m) aboveground in Exposure C: (a) western United States; and (b) eastern United States. (Continued)

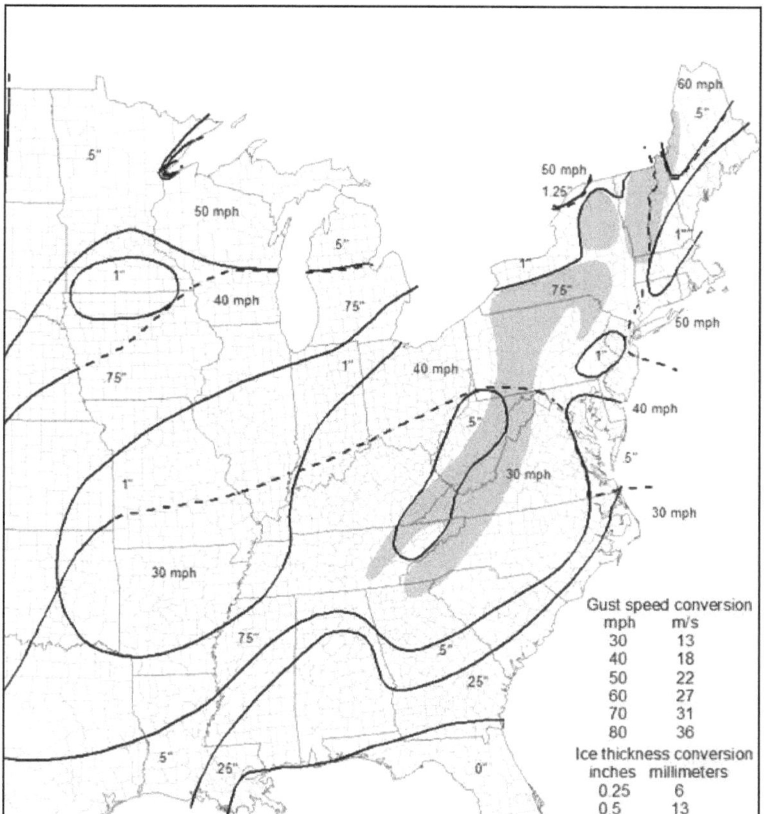

(b)

Figure L-2. (Continued) 50-year MRI radial ice thickness (in.) from freezing rain with concurrent gust wind speeds (mph) at 33 ft (10 m) aboveground in Exposure C: (a) western United States; and (b) eastern United States.

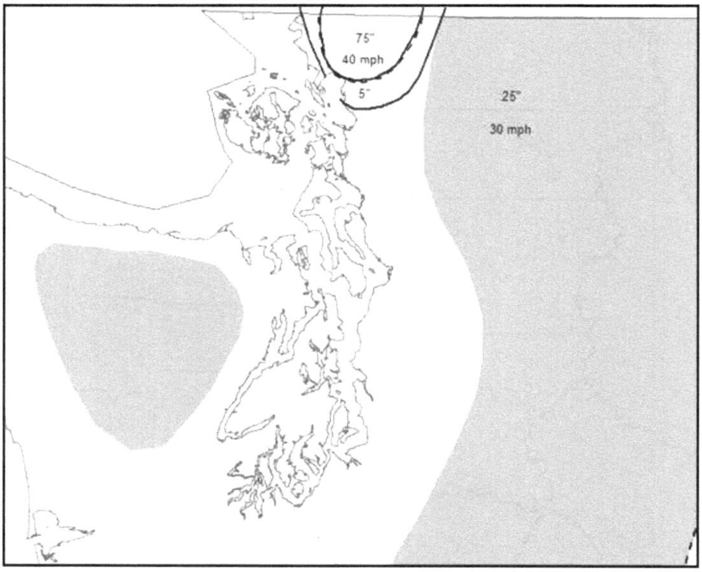

Figure L-3. 50-year MRI radial ice thickness (in.) from freezing rain with concurrent gust wind speeds (mph) at 33 ft (10 m) aboveground in Exposure C: Puget Sound detail.

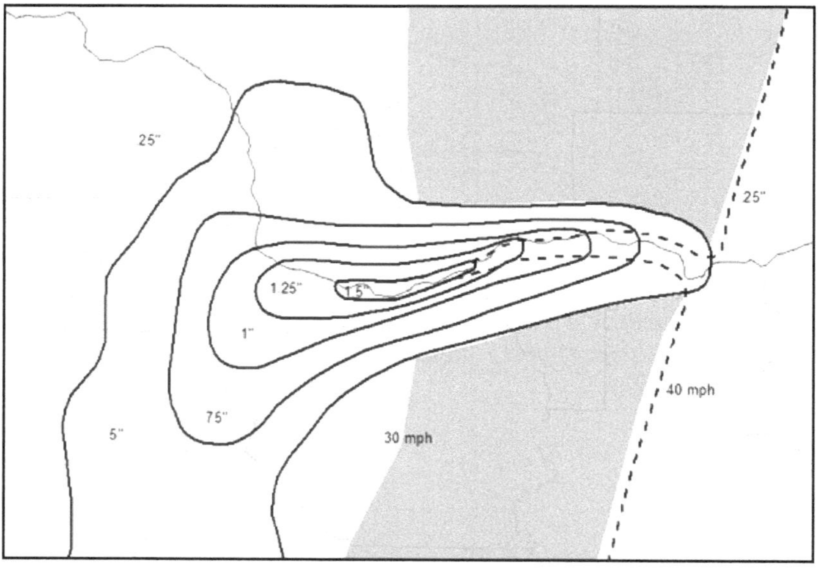

Figure L-4. 50-year MRI radial ice thickness (in.) from freezing rain with concurrent gust wind speeds (mph) at 33 ft (10 m) aboveground in Exposure C: Columbia River Gorge detail.

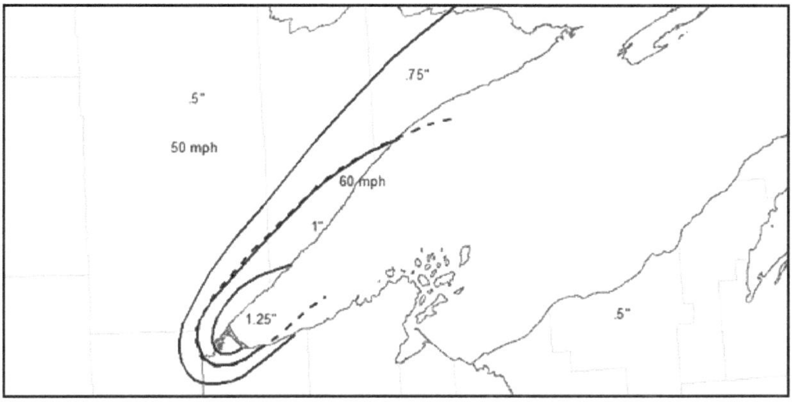

Figure L-5. 50-year MRI radial ice thickness (in.) from freezing rain with concurrent gust wind speeds (mph) at 33 ft (10 m) aboveground in Exposure C: Lake Superior detail.

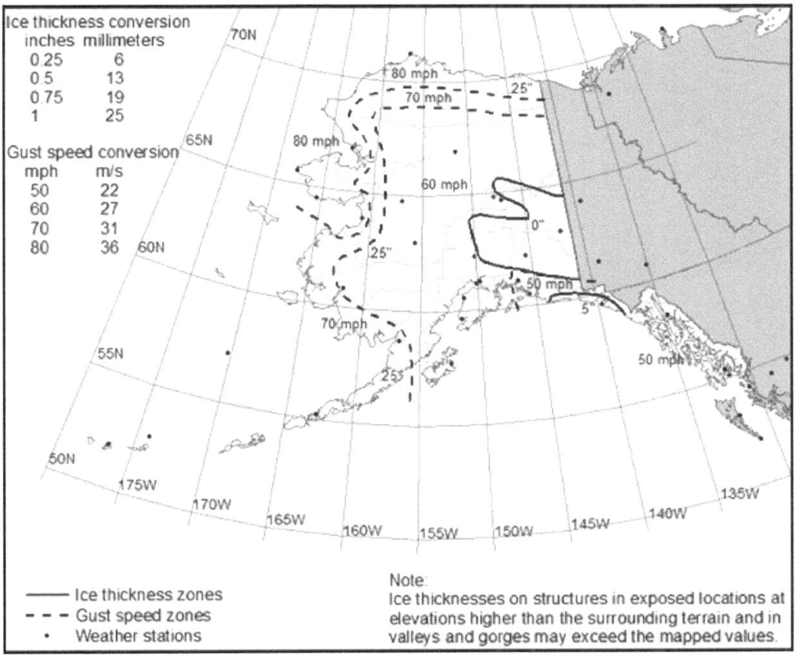

Figure L-6. 50-year MRI radial ice thickness (in.) from freezing rain with concurrent gust wind speeds (mph) at 33 ft (10 m) aboveground in Exposure C: Alaska.

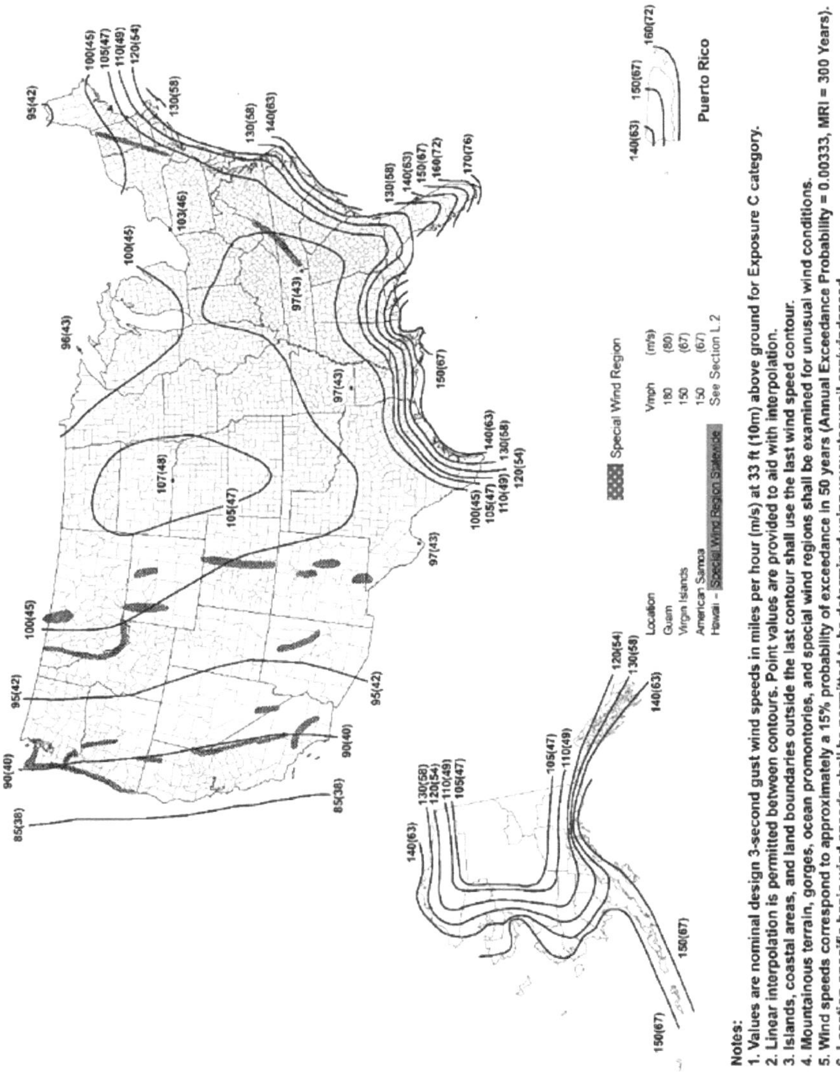

Figure L-7. 300-year MRI 3-second gust wind speed map [mph (m/s)] at 33 ft (10 m) aboveground in Exposure Category C.
Source: ASCE (2017).

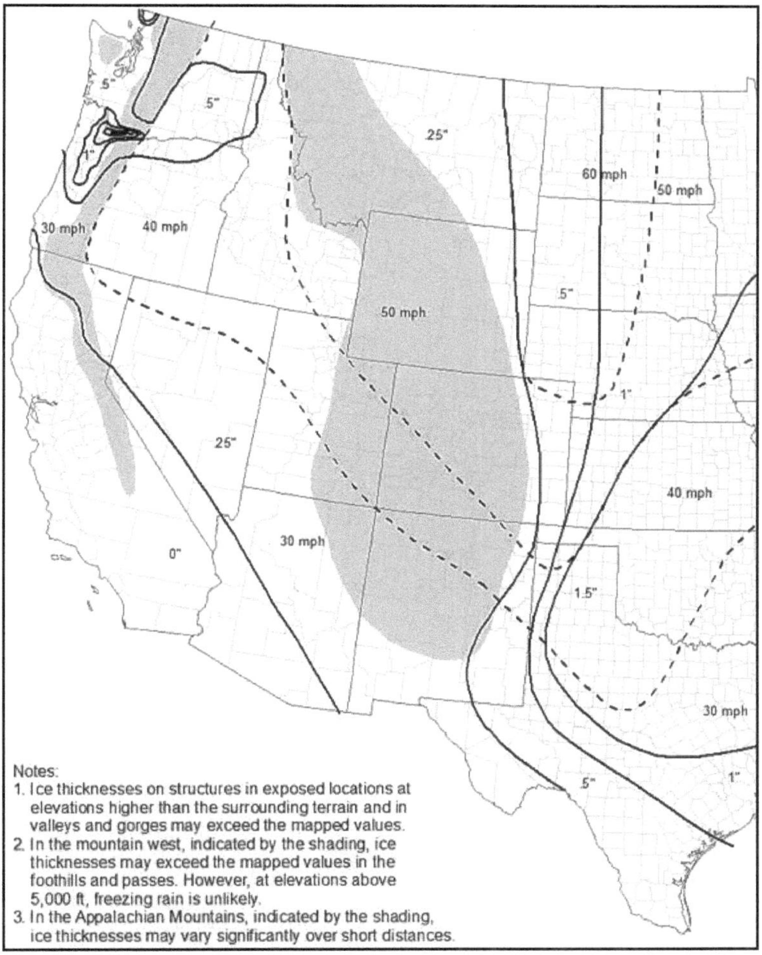

(a)

Figure L-8. 300-year MRI radial ice thickness (in.) from freezing rain with concurrent gust wind speeds (mph) at 33 ft (10 m) aboveground in Exposure C: (a) western United States; and (b) eastern United States. (Continued)

(b)

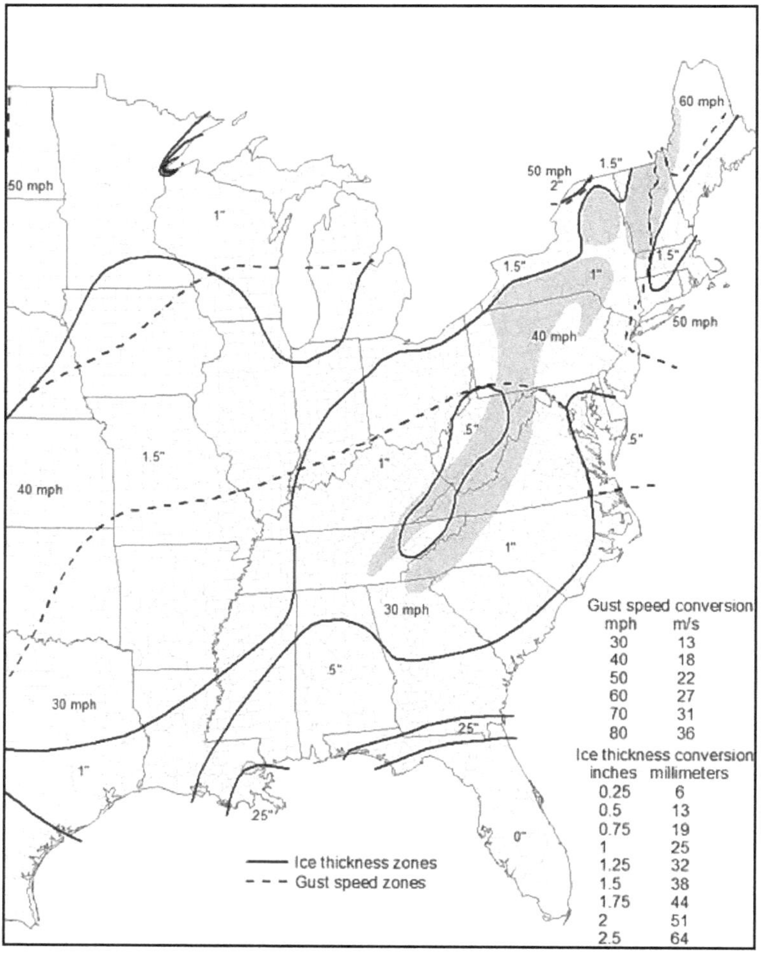

Figure L-8. (Continued) 300-year MRI radial ice thickness (in.) from freezing rain with concurrent gust wind speeds (mph) at 33 ft (10 m) aboveground in Exposure C: (a) western United States; and (b) eastern United States.

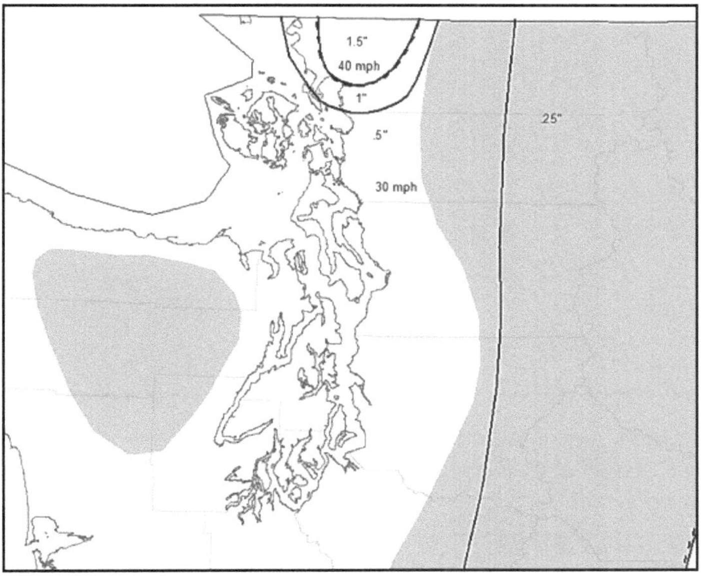

Figure L-9. 300-year MRI radial ice thickness (in.) from freezing rain with concurrent gust wind speeds (mph) at 33 ft (10 m) aboveground in Exposure C: Puget Sound detail.

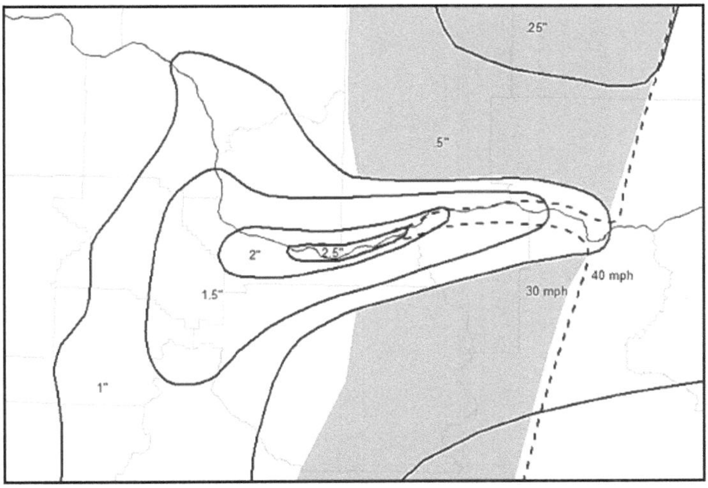

Figure L-10. 300-year MRI radial ice thickness (in.) from freezing rain with concurrent gust wind speeds (mph) at 33 ft (10 m) aboveground in Exposure C: Columbia River Gorge detail.

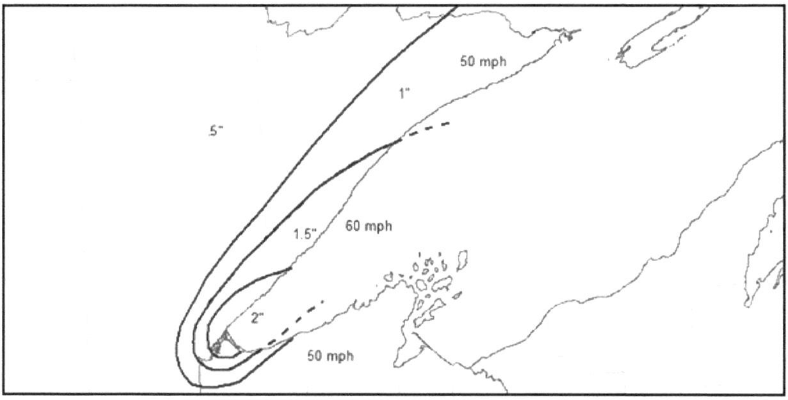

Figure L-11. 300-year MRI radial ice thickness (in.) from freezing rain with concurrent gust wind speeds (mph) at 33 ft (10 m) aboveground in Exposure C: Lake Superior detail.

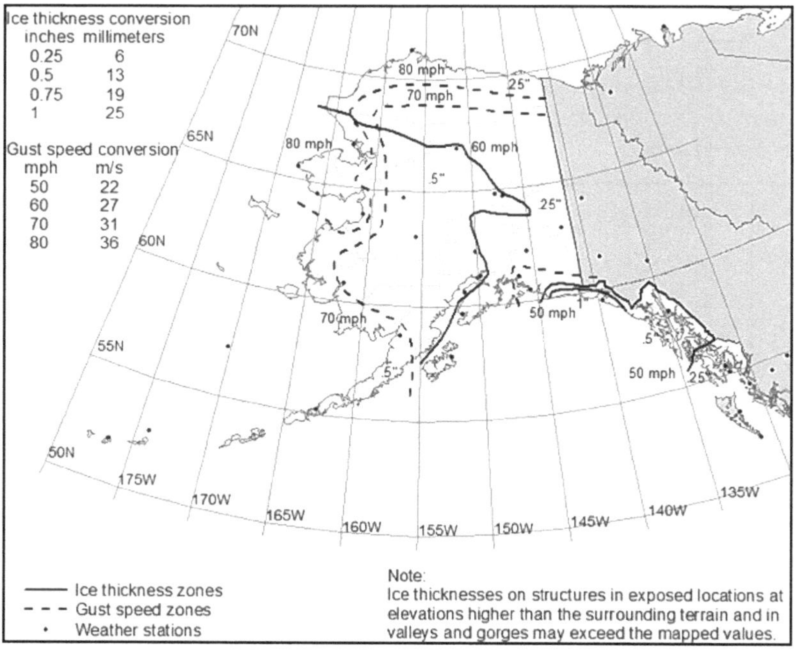

Figure L-12. 300-year MRI radial ice thickness (in.) from freezing rain with concurrent gust wind speeds (mph) at 33 ft (10 m) aboveground in Exposure C: Alaska.

APPENDIX M

DRAFT PRE-STANDARD MINIMUM DESIGN LOADS FOR ELECTRICAL TRANSMISSION LINE FACILITIES

M.1 PURPOSE

The ASCE Task Committee on Structural Loadings envisions the need for a loading standard for transmission line facilities. This appendix presents the recommendations of this committee written in a prescriptive form and presented as a Draft Pre-Standard for public review and comment. Comments should be directed to the ASCE Committee on Electrical Transmission Structures.

M.2 SCOPE

This Draft Pre-Standard provides minimum load requirements for the design of transmission line facilities. Appropriate loads, load factors, and loading combinations that have been developed to be used concurrently are set forth for ultimate strength design (LRFD) using the referenced design standards and guides.

This Draft Pre-Standard shall be used by the Transmission Owner or their authorized agent responsible for developing the structural load cases for the design of transmission line facilities. The minimum voltage that applies to a transmission line shall be established by the Transmission Owner.

NOTE: This Draft Pre-Standard is intended for design of new transmission line facilities. It may be applied to the assessment of existing facilities. Further, these principles may be applied to temporary or emergency facilities with adjustments.

NOTE: These principles may be applied to distribution facilities.

M.3 APPLICABLE DOCUMENTS

The following standards, codes, and guidelines are referenced in this Draft Pre-Standard; the latest revisions apply unless noted:

- ACI 318, *Building Code Requirements for Structural Concrete and Commentary*;
- AISC 360, *Specification for Steel Buildings*;
- ANSI C2, *National Electrical Safety Code* (NESC);
- ANSI C29, Series of standards for both ceramic and non-ceramic insulators;
- ANSI C119, Series of standards for conductors;
- ANSI O5, Series of standards for wood products;
- ASCE 7-16, *Minimum Design Loads and Associated Criteria for Buildings and Other Structures*;
- ASCE 10, *Design of Latticed Steel Transmission Structures*;
- ASCE 48, *Design of Steel Transmission Pole Structures*;
- ASCE MOP 74, *Guidelines for Electrical Transmission Line Structural Loading*;
- ASCE MOP 91, *Design of Guyed Electrical Transmission Structures*;
- ASCE MOP 104, *Recommended Practice for Fiber-Reinforced Polymer Products for Overhead Utility Line Structures*;
- ASCE MOP 123, *Prestressed Concrete Transmission Pole Structures*;
- ASCE MOP 141, *Wood Pole Structures for Electrical Transmission Lines*;
- ASCE *Guide for the Design and Use of Concrete Poles*;
- IEEE Standard 524, *IEEE Guidelines to the Installation of Overhead Transmission Line Conductors*;
- IEEE Standard 951, *IEEE Guide to the Assembly and Erection of Metal Transmission Structures*;
- IEEE Standard 1025, *IEEE Guide to the Assembly and Erection of Concrete Pole Structures*;

- IEEE 1307, *IEEE Standard for Fall Protection for Electric Utility Transmission and Distribution on Poles and Structures*; and
- OSHA Occupational Safety and Health Administration applicable regulations.

M.4 DEFINITIONS

Basic Wind Speed: The 3-second gust wind speed at 33 ft (10 m) above ground in open country terrain (Exposure Category C).

Effective Height: The theoretical height above ground to the center of pressure of the wind load.

Exposure Category: A description of the terrain features and ground roughness upwind of the transmission line facility.

Force Coefficient: A coefficient accounting for the effects of member characteristics (e.g., shape, size, solidity, shielding, orientation with respect to the wind, surface roughness) on the resultant force due to wind. It is also referred to as the drag coefficient, pressure coefficient, or shape factor.

Freezing Rain: Rain or drizzle that falls into a layer of subfreezing air at the earth's surface and freezes on contact with the ground or an object to form glaze ice.

Glaze Ice: Clear high-density ice, with a density of approximately 56 pcf (900 kg/m^3).

Gust Response Factor: The ratio of the peak load effect on the structure or wires to the mean load effect corresponding to the design wind speed. It is a multiplier of the design wind load to obtain the peak load effect.

Hoarfrost: An accumulation of ice crystals formed by direct deposition of water vapor from the air onto an object.

Ice Accretion: The formation of ice on transmission line facilities and nonstructural attachments.

In-Cloud Icing: Ice occurring when supercooled cloud or fog droplets carried by the wind freeze on impact with objects. In-cloud icing usually forms rime, but may also form glaze.

Longitudinal: Local axis of the structure that is generally parallel to the direction of the wires and perpendicular to the vertical axis.

Mean Recurrence Interval (MRI): The inverse of the probability of exceedance of an environmental load (i.e., wind, ice) in any given year. For example, a design event with the probability of exceedance of 0.01 (1%) in any given year is associated with an MRI of 100 years.

Nonstructural Attachments: Components attached to the structure or wires with mass or surface area that significantly contributes to the overall structural loading. Such attachments include, but are not limited to, electrical and communication equipment, signs, ladders, platforms, aerial marker balls, bird diverters, and anti-galloping devices.

Rime Ice: White or opaque ice with entrapped air. Typical densities range from 20 to 50 pcf (320 to 800 kg/m^3).

Snow: Snow that adheres to objects by some combination of capillary forces, freezing, and sintering. The snow may be either wet or dry. Typical densities range from 20 to 60 pcf (320 to 960 kg/m^3).

Solidity Ratio: A ratio of the area of all members in the windward face of a latticed structure (net area) to the area of the outline of the windward face of a latticed structure (gross area).

Transmission Line Facilities: All wires, insulators, hardware, supporting structures, guy wires, anchors, and foundations.

Transverse: Local axis of the structure that is generally perpendicular to the direction of the wires and the vertical axis.

Vertical: Local axis of the structure that is parallel with the direction of the gravitational force.

Wind Pressure Exposure Coefficient: A coefficient used to modify the basic wind pressure to account for variations of wind speed with height due to interaction (friction) with the surface roughness of the earth.

Wire: All electrical conductors, shield wires, optical ground wire, messengers, and communication cables attached to a transmission structure.

Yawed Wind: Wind at angles of incidence to the transmission structure or line other than the longitudinal and transverse loading directions.

M.5 SYMBOLS

The following symbols are used in this Draft Pre-Standard:

A = Area projected on a plane normal to the wind direction [ft^2 (m^2)];

A_m = Area of all members in the windward face of a latticed structure (net area) [ft^2 (m^2)];

A_o = Area of the outline of the windward face of a latticed structure (gross area) [ft^2 (m^2)];

B_t = Dimensionless response term corresponding to the quasi-static background wind loading on the structure;

B_w = Dimensionless response term corresponding to the quasi-static background wind loading on the wires;

c_{exp} = Turbulence intensity constant, based on exposure;

C_f = Force coefficient associated with the windward face of the structure;

d = Diameter of bare wire [inches (mm)];

F = Wind force in the direction of wind unless otherwise specified [lb (N)];

G = Gust response factor for structures and wires;

G_t = Gust response factor for the structure;
G_w = Gust response factor for the wires;
HIW = High Intensity Wind
I_z = Turbulence intensity at the effective height of the tower/structure or wire;
K_z = Wind pressure exposure coefficient, which modifies the reference wind pressure for various heights above ground based on different exposure categories;
K_{zt} = Topographic factor;
LC = Load Case
L_s = Integral length scale of turbulence [ft (m)];
MRI = Mean Recurrence Interval
Q = Air density coefficient in the wind force equation that converts the kinetic energy of moving air into potential energy of pressure;
Q_D = Design load effect in each component of a structure;
Q_i = Constant to convert radial ice thickness to weight;
Q_{MRI} = Reliability adjustment factor;
S = Design wind span [ft (m)] of the wires;
t = Nominal ice thickness due to freezing rain at a height of 33 ft (10 m) [inches (mm)];
t_{MRI} = Nominal ice thickness due to freezing rain at a height of 33 ft (10 m) at a selected MRI [inches (mm)];
t_z = Design ice thickness [inches (mm)];
t_{100} = Nominal ice thickness attributable to freezing rain at a height of 33 ft (10 m) for 100-year MRI from Figures M-2 through M-7 [inches (mm)];
V_{MRI} = Reference 3-second gust wind speed for selected MRI [mph (m/s)];
V_{100} = Reference 3-second gust wind speed for 100-year MRI from Figure M-1 [mph (m/s)];
w = Wire weight per unit length [lb/ft (N/m)];
W_i = Weight of glaze ice per unit length [lb/ft (N/m)];
z = Height aboveground [ft (m)];
z_g = Gradient height, which defines the thickness of the atmospheric boundary layer [ft (m)]; above this height, the wind speed is constant;
z_h = Effective height from ground level to the center of pressure of the wind load on the structure, or effective height of the wire [ft (m)];
α = Power law coefficient for gust wind;
γ = Load factor appropriate for the event;
Φ = Solidity ratio (A_m/A_o);
$\varphi \cdot R_n$ = Design resistance or deflection restriction of the transmission line facility; and
Ψ = Yaw angle measured in a horizontal plane, referenced as 0° for wind perpendicular to the wires (degrees).

M.6 LOAD CASES FOR STRENGTH DESIGN

The Transmission Owner or authorized agent shall design transmission line facilities with sufficient strength for the basic load cases defined in Section M.6.1. Consideration shall also be given to the supplemental and serviceability load cases defined in Section M.6.2.

Load cases shall be multiplied by the applicable load factors defined in Section M.6.3.

Where transmission line facilities warrant a reliability level different from that defined in this Draft Pre-Standard (i.e., MRI100) due to site-specific application, the provisions of Section M.6.4 shall be followed.

The load cases of this section are cumulatively represented by the following:

$$\Sigma\,(\gamma \cdot LC \cdot Q_{MRI}) \tag{M-1}$$

where

LC = Load Case defined in Sections M.6.1 and M.6.2,

γ = Load factor appropriate for the event defined in Section M.6.3, and

Q_{MRI} = Reliability adjustment factor defined in Section M.6.4.

M.6.1 Basic Load Cases

Transmission line facilities shall be designed such that their design strength equals or exceeds the effects of the following load cases. Determination of the magnitude of each load, including the effects of dead and wire loads, shall be in accordance with the applicable section of this Draft Pre-Standard.

- Climatic Load Cases
  - Extreme wind load
  - Extreme ice load with concurrent wind load and temperature
- Line Security Load Cases
  - Unbalanced longitudinal load
  - Failure containment load
- Operational Load Cases
  - Construction load with concurrent weather condition
  - Maintenance load with concurrent weather condition
  - Equipment operation load when applicable
- Legislated Loads

M.6.2 Supplemental and Serviceability Load Cases

Where site-specific circumstances warrant, the following load cases shall be considered in the design of transmission line facilities. These loads may not have an associated MRI:

- Load criteria associated with deflection limitations.
- Dynamic wire loading with associated weather event.
  NOTE: Dynamic loading should be evaluated relative to appropriate component resistance considering the nature of the loading.
- Earthquake events
  NOTE: The structural capacity provided by designing transmission line facilities to the loading requirements of this Draft Pre-Standard provides sufficient capability to resist earthquake ground motions (inertia loads). Site and soil conditions and foundation types should be reviewed for projects in high seismic regions to address potential failures from secondary events such as soil liquefaction and landslides.
- Other potential high consequence events such as floods, tsunamis, snow creep, and avalanches. See Section M.13 for additional information.

M.6.3 Load Factors

Unless otherwise specified within this Draft Pre-Standard, the following load factors (γ) shall be the minimum used. Any unique situation deemed by the Transmission Owner or authorized agent to warrant a load factor greater than 1.0 shall be considered.

- 1.0: All weather-related loads.
  Note: ASCE Manual of Practice No. 74 prefers designers use MRI to adjust resistance to weather events based on the importance of specific facility. See Section M.6.4.
- 1.5: Construction and maintenance loads.
- 1.0: Dead load, structure weight, and weight of supported facilities.
- Legislated loads: Use the load factors as defined by the legislated document.

M.6.4 Reliability Adjustment

The MRI used in the calculation of climatic loads may be adjusted for transmission line facilities requiring a reliability level different from that defined within this Draft Pre-Standard (i.e., MRI_{100}). This may be due to site-specific applications or studies, or operating circumstances such as those described below:

- Challenging site access and restoration circumstances
  Examples: River, highway, railroad crossing;

- Temporary construction;
- Emergency restoration; and
- The importance of the line relative to the performance of the transmission grid or service provided.

The Transmission Owner or authorized agent shall determine the appropriate MRI for those situations warranting one different from that specified as the baseline of this Draft Pre-Standard. Wind speed and ice accretion maps for other MRIs can be found in Appendix L of ASCE Manual of Practice No. 74 or ASCE 7-16.

Note: Climatic load cases as defined in this Draft Pre-Standard are determined by statistical modeling techniques. As such, the structural reliability level can be adjusted.

M.6.5 Strength

The resistance of a transmission line facility shall exceed the effects of the prescribed loads in this document as described by the following formula:

$$\varphi \cdot R_n \geq \Sigma(\gamma \cdot \mathrm{LC} \cdot Q_{\mathrm{MRI}}) \tag{M-2}$$

where

$\varphi \cdot R_n$ = design resistance or deflection restriction of the transmission line facility as defined by the appropriate design guide for the applicable structure type or an appropriate serviceability restriction.

The strength of transmission line facilities is beyond the scope of this Draft Pre-Standard. The Transmission Owner or authorized agent is referred to the following documents for the applicable material strength factors and design requirements:

- ACI 318, *Building Code Requirements for Structural Concrete and Commentary;*
- AISC 360, *Specifications for Steel Buildings;*
- ANSI C2, *National Electrical Safety Code;*
- ANSI C29, Series of standards for both ceramic and non-ceramic insulators;
- ANSI C119, Series of standards for conductors;
- ANSI O5, Series of standards for wood;
- ASCE Standard No. 10, *Design of Latticed Steel Transmission Structures;*
- ASCE Standard No. 48, *Design of Steel Transmission Pole Structures;*
- ASCE Manual of Practice No. 91, *Design of Guyed Electrical Transmission Structures;*

- ASCE MOP 104, *Recommended Practice for Fiber-Reinforced Polymer Products for Overhead Utility Line Structures;*
- ASCE MOP 123, *Prestressed Concrete Transmission Pole Structures*; and
- ASCE MOP 141, *Wood Pole Structures for Electrical Transmission Lines.*

For strengths of other transmission line facilities such as insulators and conductors, refer to the specifications of the applicable supplier.

M.7 DEAD LOADS

The weight of the structure, wires, and components such as insulators, hardware, electrical equipment, and non-structural attachments shall be included in the design of transmission line structures and foundations.

M.8 WIRE LOADS

M.8.1 General

The loads induced by all attached wires known at the time of initial design shall be included in the design of the structure. Wire loads shall be calculated based on tensions, span lengths, and line angles appropriate for the site and for the temperature, ice, and wind loadings specified in this Draft Pre-Standard. The effects of wind and ice on non-structural attachments attached to wires shall be used in the calculation of design wire tensions. Further, wire loads shall be applied in the various combinations defined in Section M.6. Climatic, construction, and legislated loads shall be included and combined as appropriate to determine the maximum load effect.

M.8.2 Dynamic Wire Loads

Dynamic wire loads, such as those resulting from galloping, ice shedding, and aeolian vibration that are caused or enhanced by wind and ice or flexible structures and supports, shall be considered.

M.8.3 Unbalanced Longitudinal Loads

Unbalanced longitudinal loads resulting from unequal tensions on adjacent spans due to variation in icing or wind speed; broken shield wire, conductor, or sub-conductors; conductor stringing; and secondary loading resulting from a structural failure shall be considered.

M.9 WIND LOADS

M.9.1 General

Transmission line facilities shall be designed to resist the wind loads determined in accordance with this section. Wind loading with all other applicable loads addressed herein shall be applied in the direction that produces the maximum load effect.

M.9.2 Wind Force

The wind force acting on the projected surface area of components of transmission line facilities and non-structural attachments shall be determined by Equation (M-3a) or (M-3b)

$$F = QK_zK_{zt}(V_{100})^2GC_fA \tag{M-3a}$$

or

$$F = QK_zK_{zt}(V_{\mathrm{MRI}})^2GC_fA \tag{M-3b}$$

where

- F = Wind force in the direction of wind unless otherwise specified [lb (N)];
- G = Gust response factor for structures and wires as specified in Section M.9.2.6;
- C_f = Force coefficient as defined in Section M.9.2.7;
- A = Area of the component projected on the plane normal to the wind direction [ft2 (m2)];
- Q = Air density coefficient defined in Section M.9.2.1;
- K_z = Wind pressure exposure coefficient which modifies the reference wind pressure for various heights above ground based on different exposure categories; the values are obtained from Section M.9.2.5;
- K_{zt} = Topographic factor; 1.0 unless the guidance in Section M.9.2.8 and the procedures of ASCE 7 are followed;
- V_{100} = Reference 3-second gust wind speed for 100-year MRI [mph (m/s)] obtained from Figure M-1 in Section M.9.2.2; and
- V_{MRI} = Reference 3-second gust wind speed for selected MRI [mph (m/s)] obtained from ASCE 7-16 or Appendix L of ASCE Manual of Practice 74.

M.9.2.1 Air Density Coefficient, Q For wind speed in miles per hour (m/s) and pressure in pounds per square foot (Pa), Q is defined in Equation (M-4). A different value of Q may be used if justified by analysis of site-specific elevation and temperature data. Refer to ASCE Manual of Practice No. 74, Appendix C for additional information.

$$Q = 0.00256 \text{ customary units } (0.613 \text{ metric units}) \qquad \text{(M-4)}$$

M.9.2.2 Basic Wind Speed The basic wind speed associated with a 100-year MRI, V_{100}, used in the determination of design wind loads on transmission line facilities and non-structural attachments shall be determined from Figure M-1, except as provided in Sections M.9.2.2.1 and M.9.2.2.2. Linear interpolation between contours is permitted. The last wind speed contour of the coastal area may be used for islands and coastal areas beyond the last contour.

If wind speeds associated with a MRI other than 100 years are required, the designer is referred to the additional wind speed maps in Appendix L of ASCE Manual of Practice No. 74 or ASCE 7-16.

M.9.2.2.1 Special Wind Regions Mountainous terrain, gorges, or other special wind regions shown in Figure M-1 shall be examined for unusual wind conditions. If necessary, the designer shall adjust the values given in Figure M-1 to account for local wind speeds. Such adjustment shall be based on meteorological information and an estimate of the basic wind speed obtained in accordance with the provisions of Section M.9.2.2.2.

For transmission line facilities in the state of Hawaii, the basic wind speed shall be determined as follows:

$$105 \text{ mph } (47 \text{ m/s}) \text{ for regions indicated as } K_{zt} \leq 1.5$$

$$\left(86\,\text{mph} \bullet \sqrt{K_{zt}}\right) \text{ or } \left(38\,\text{m/s} \bullet \sqrt{K_{zt}}\right) \text{ for regions indicated as } K_{zt} > 1.5$$

Values of K_{zt} shall be determined using the appropriate wind map available from the Department of Accounting and General Services for the state of Hawaii.

M.9.2.2.2 Estimation of Basic Wind Speeds from Regional Climatic Data In areas outside hurricane-prone regions, regional climatic data shall only be used in lieu of the basic wind speeds given in Figure M-1 when (1) approved extreme value statistical analysis procedures have been employed in reducing the data; and (2) the length of record, sampling error, averaging time, anemometer height, data quality, and terrain exposure of the anemometer have been taken into account. The extreme value statistical analysis may be used to justify a reduction in basic wind speed below that of Figure M-1.

The use of regional wind speed data obtained from anemometers is not permitted to define the hurricane wind speed risk along the hurricane-prone regions of the continental United States, Hawaii, Puerto Rico, Guam, the Virgin Islands, and American Samoa. In hurricane-prone regions, wind speeds derived from simulation techniques shall only be used in lieu of the basic wind speeds given in Figure M-1 when

1. Wind industry accepted simulation procedures are applied (i.e., Monte Carlo simulations based on historical hurricane records).
2. An appropriate number of years of synthetic hurricane activity are simulated and validated using historical key hurricane statistics. NOTE: A minimum database of 50,000 to 100,000 years of simulated hurricane activity is typically considered acceptable.
3. A wind engineering industry-accepted wind field model is used to generate wind speeds based on hurricane track records.
4. Wind industry accepted extreme value statistical analysis procedures are used for the estimation of extreme wind speeds (i.e., Type I extreme value distribution) at a given location.

In areas outside hurricane-prone regions, when the basic wind speed is estimated from regional climatic data, the basic wind speed shall not be less than the wind speed associated with the specified mean recurrence interval, and the estimate shall be adjusted for equivalence to a 3-second gust wind speed at 33 ft (10 m) above ground in Exposure Category C.

M.9.2.3 Limitations The load effects resulting from localized convective winds, also referred to as high intensity winds (HIWs), such as thunderstorms, downbursts, and tornadoes, are not addressed by this Draft Pre-Standard. Refer to ASCE Manual of Practice No. 74 for additional information related to HIW.

M.9.2.4 Exposure Categories Exposure Category C shall be used to design transmission facilities, unless the criteria of Exposures B or D can be met and the exposure category will not change over the life of the transmission line.

Exposure Category B is classified as urban and suburban areas, densely wooded areas, or terrain with numerous, closely spaced obstructions having the size of single-family dwellings or larger. Use of Exposure Category B shall be limited to wind directions for which representative terrain extends either 2,600 feet (792 m) or 20 times the height of the transmission structure, whichever is greater.

In the use of Exposure Category B, the longest distance of flat, unobstructed terrain located in the middle of a suburban area permitted before Exposure Category C must be used is 600 ft (180 m) or 20 times the height of the transmission structure, whichever is less.

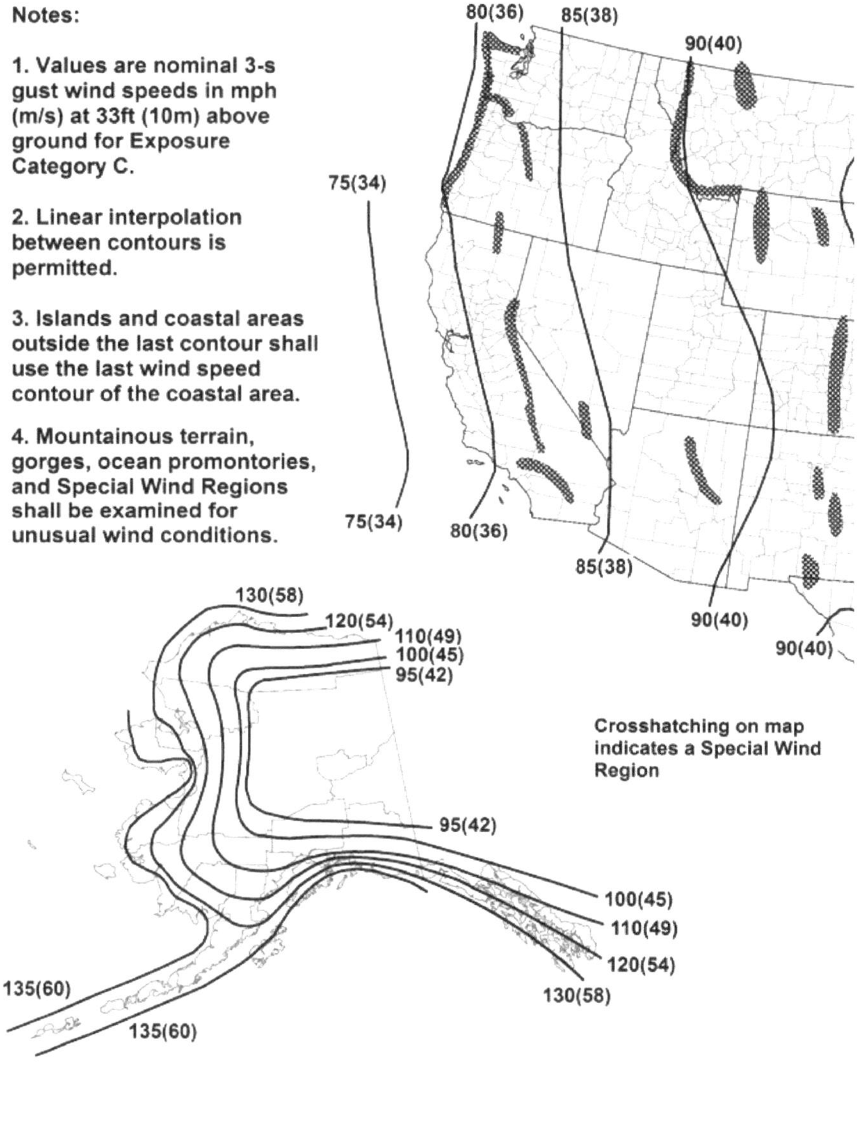
Notes:
1. Values are nominal 3-s gust wind speeds in mph (m/s) at 33ft (10m) above ground for Exposure Category C.
2. Linear interpolation between contours is permitted.
3. Islands and coastal areas outside the last contour shall use the last wind speed contour of the coastal area.
4. Mountainous terrain, gorges, ocean promontories, and Special Wind Regions shall be examined for unusual wind conditions.
80(36)
85(38)
90(40)
75(34)
75(34)
80(36)
85(38)
90(40)
90(40)
130(58)
120(54)
110(49)
100(45)
95(42)
Crosshatching on map indicates a Special Wind Region
95(42)
100(45)
110(49)
120(54)
130(58)
135(60)
135(60)

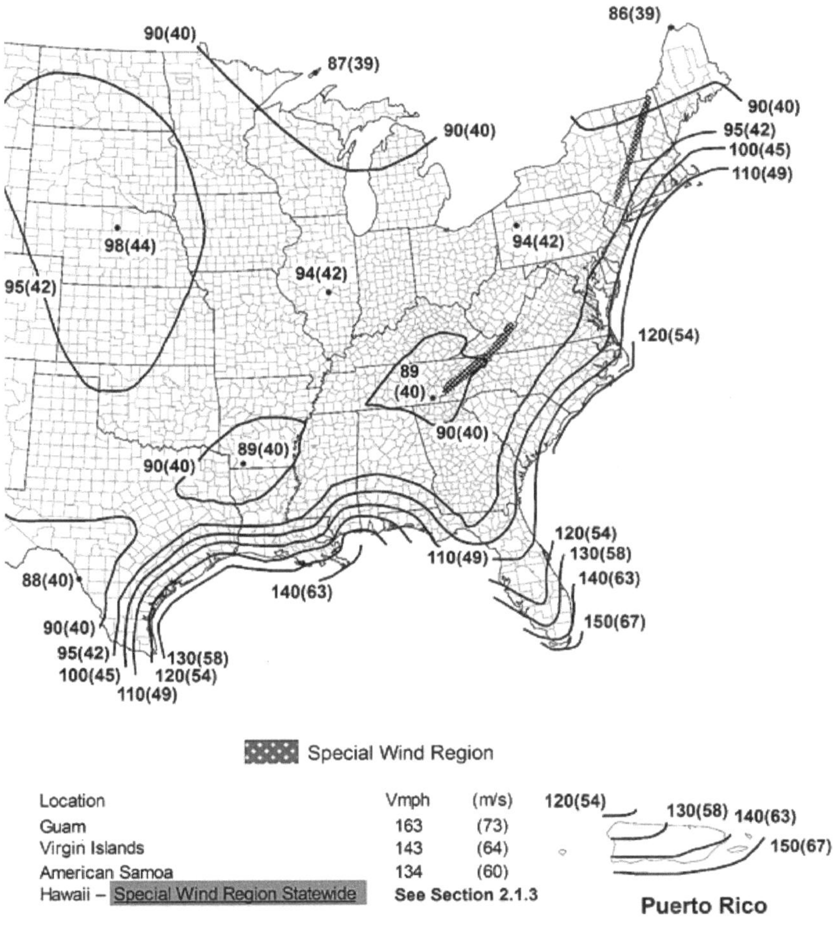

Figure M-1. 100-year MRI 3-second gust wind speed map [mph (m/s)] at 33 ft (10 m) above ground in Exposure C.
Source: ASCE (2017).

Exposure Category C is defined as open terrain with scattered obstructions having heights less than 30 ft (9.1 m). This category includes flat, open country, farms, and grasslands. This exposure category should be used whenever terrain does not fit the descriptions of the other exposure categories. This exposure category may be considered representative of airport terrain, where most wind speed measurements are recorded.

Exposure Category D is described as flat, unobstructed areas directly exposed to wind flowing over open water for a distance of at least 5,000 ft (1,524 m), or 20 times the height of the structure, whichever is greater. This exposure category applies to shorelines in hurricane-prone regions, inland waterways, the Great Lakes, and coastal areas of California, Oregon, Washington, and Alaska. This exposure category applies only to structures directly exposed to bodies of water and coastal beaches. Exposure Category D extends inland from the shoreline a distance of 1500 ft (457 m) or 10 times the height of the transmission structure, whichever is greater.

For a site located in the transition zone between exposure categories, or where the exposure category is determined to be different on opposite sides of the transmission facility, the category resulting in the largest wind forces shall be used.

Exposure Categories for the state of Hawaii shall be determined using the appropriate wind map available from the Department of Accounting and General Services for the state of Hawaii.

NOTE: For additional information related to the definition of Exposure Categories and how to apply them to facilities in the transition zones between categories, refer to Chapter 2 of ASCE Manual of Practice No. 74.

M.9.2.5 Wind Pressure Exposure Coefficient

The wind pressure exposure coefficient, K_z, used in Equation (M-1) shall be calculated using Equation (M-5)

$$K_z = 2.01\left(\frac{z_h}{z_g}\right)^{\frac{2}{\alpha}} \quad \text{for } 33 \text{ ft} \le z_h \le z_g \tag{M-5}$$

where

α = Power law coefficient for gust wind from Table M.9-1
z_h = Effective height from ground level to the center of pressure of the wind load, and
z_g = Gradient height from Table M.9-1

For heights up to 200 ft (60 m) from ground level, K_z may be determined using the values shown in Table M.9-2.

NOTE: For structural design purposes, the effective heights of all wires may be approximated as the average height above ground of all the wire attachment points to the structure. For structure heights 200 ft (60 m) or less, the design engineer may assume the structure is comprised of one section, and two-thirds of the total structure height may be used as the effective height.

Table M.9.1. Power Law Exponent for Gust Wind Speed and Corresponding Gradient Height

Exposure	α	z_g (ft)
B	7.0	1200
C	9.5	900
D	11.5	700

Table M.9-2. Wind Pressure Exposure Coefficient, K_z

Effective height, z_h (ft)	Wind pressure exposure coefficient, K_z		
	Exposure B	Exposure C	Exposure D
0–33	0.72	1.00	1.18
40	0.76	1.04	1.22
50	0.81	1.09	1.27
60	0.85	1.14	1.31
70	0.89	1.17	1.35
80	0.93	1.21	1.38
90	0.96	1.24	1.41
100	0.99	1.27	1.43
120	1.04	1.32	1.48
140	1.09	1.36	1.52
160	1.13	1.40	1.55
180	1.17	1.43	1.59
200	1.20	1.46	1.62

M.9.2.6 Gust Response Factor

Structure and wire gust response factors, G_t and G_w, respectively, shall be determined using Equations (M-6) and (M-7).

$$G_t = \left(\frac{1 + 4.6 I_z B_t}{1 + 6.1 I_z} \right) \quad \text{(M-6)}$$

$$G_w = \left(\frac{1 + 4.6 I_z B_w}{1 + 6.1 I_z} \right) \quad \text{(M-7)}$$

in which

$$I_z = c_{\exp} \left(\frac{33}{z_h} \right)^{\frac{1}{6}} \quad \text{(M-8)}$$

$$B_t = \sqrt{\frac{1}{1 + \frac{0.56 z_h}{L_s}}} \quad \text{(M-9)}$$

$$B_w = \sqrt{\frac{1}{1 + \frac{0.8S}{L_s}}} \quad \text{(M-10)}$$

where

B_t = Dimensionless response term corresponding to the quasi-static background wind loading on the structure,
B_w = Dimensionless response term corresponding to the quasi-static background wind loading on the wires,
$c_{\exp}$ = Turbulence intensity constant based on exposure and found in Table M.9-3,
I_z = Turbulence intensity at effective height of the tower/structure or wire,
L_s = Integral length scale of turbulence [ft (m)] found in Table M.9-3,
S = Design wind span [ft (m)] of the wires, and
z_h = Two-thirds of the total height of the structure for the calculation of G_t using Equation (M-6), or effective height of the wire for the calculation of G_w using Equation (M-7).

Table M.9-3. Turbulence Parameters for Calculation of Gust Response Factor by Exposure

Exposure	$c_{\exp}$	L_s (ft)
B	0.3	170
C	0.2	220
D	0.15	250

The complete gust response factor equations in Appendix F of ASCE Manual of Practice No. 74 shall be considered for the calculation of G_t and G_w when

G_t = Structure natural frequency is less than 1 Hz, (which typically occurs with structures taller than 200 ft);
G_w = Horizontal span is longer than 2000 ft.

M.9.2.7 Force Coefficient For members with aspect ratios greater than or equal to 40, the force coefficient shall be determined from Sections M.9.2.7.1 through M.9.2.7.3. For members with aspect ratios less than 40, the correction factors defined in Appendix G of ASCE Manual of Practice No. 74 shall be applied.

For force coefficients of shapes not described within this Draft Pre-Standard, refer to Appendix G of the ASCE Manual of Practice No. 74.

M.9.2.7.1 Wires The force coefficient for single and bundled conductors and for ground wires shall be as defined in Equation (M-11) unless more definitive data based on wind force measurements are available.

$$C_f = 1.0 \qquad \text{(M-11)}$$

M.9.2.7.2 Latticed Truss Structures Force coefficients for square-section and triangular-section latticed truss structures having flat-sided members shall be determined using Table M.9-4.

Table M.9-4. Force Coefficients, C_f, for Normal Wind on Latticed Truss Structures Having Flat-Sided Members

Tower cross section	C_f
Square	$4.0\Phi^2 - 5.9\Phi + 4.0$
Triangular	$3.4\Phi2 - 4.7\Phi + 3.4$

Source: Adapted from ASCE 7-16 (2017).

In Table M.9-4, Φ, the solidity ratio, is defined as

$$\Phi = \frac{A_m}{A_o} \qquad \text{(M-12)}$$

where

A_m = area of all members in the windward face of the structure (net area) [ft^2 (m^2)], and
A_o = area of the outline of the windward face of the structure (gross area) [ft^2 (m^2)].

When the truss members consist of round members, a correction factor, equal to $0.51\Phi 2 + 0.57$, but not greater than 1.0, shall be multiplied by the value determined in Table M.9-4.

The solidity ratio for selected panels in the transverse and longitudinal faces shall be used for the determination of wind loads.

The force coefficient calculated above accounts for loads accumulated by both the windward and leeward tower faces (including shielding effects) and shall be applied directly in Equation (M-3).

When the wind is yawed, the force coefficient C_f shall be multiplied by the wind angle magnification factor defined in Equation (M-13).

$$1 + 0.2\sin^2(2\Psi) \qquad \text{(M-13)}$$

where

Ψ = yaw angle, measured in the horizontal plane, and is referenced as 0° for wind perpendicular to the wires.

Alternatively, when wind is applied to individual members of latticed structures, force coefficients shall be 1.0 for components with round surfaces and 1.6 for components with flat surfaces. Refer to Appendix G of ASCE Manual of Practice No. 74 for values for member shapes not listed herein.

NOTE: Calculating the force coefficient using Table M.9-4 is used when applying the wind load using the "Wind on Face" method. The alternate method using the force coefficients of each individual truss member is used when applying wind load using the "Wind on Member" method.

NOTE: For latticed structures consisting of multiple columns and beams to form bent or bay structures, the force coefficients of each column or beam shall be calculated independently without considering the shielding effects of other members unless special wind studies show otherwise.

M.9.2.7.3 Pole Structures Force coefficients for pole structures are given in Table M.9-5. Refer to Appendix G of ASCE Manual of Practice No. 74 for values for member shapes not listed.

Table M.9-5. Force Coefficients, C_f, for Members of Pole Structures

Member shape	C_f	Adapted from
Circular	0.9	ASCE 7-16 (2017)
16-sided polygonal	0.9	James (1976)
12-sided polygonal	1.0	James (1976)
8-sided polygonal	1.4	ASCE 7-16 (2017), James (1976)
6-sided polygonal	1.4	ASCE 7-16 (2017)
Square, rectangle	2.0	ASCE 7-16 (2017)

M.9.2.8 Topographic Effects In complex or mountainous terrain where wind speeds vary dramatically with exposure or by channeling of wind, special studies shall be used to determine the wind flow.

For wind speed-up due to flow over isolated hills and escarpments, the procedures of ASCE Standard 7-16 shall be followed.

M.10 ICE WITH CONCURRENT WIND LOADS

M.10.1 General

Atmospheric ice loads due to freezing rain, snow, and in-cloud icing shall be considered in the design of electric transmission line facilities. Design ice thickness t_z shall be no less than the nominal ice thickness resulting from a 100-year MRI.

In areas where records or experience indicate that snow or in-cloud icing produces larger loads than from freezing rain, site-specific studies shall be used. Structural loads due to hoarfrost are not a design consideration.

M.10.1.1 Site-Specific Studies Site-specific studies shall be used to determine the 100-year MRI ice thickness, concurrent wind speed, and concurrent temperature in:

1. Alaska
2. Areas where records or experience indicate that snow or in-cloud icing produces larger loads than freezing rain.
3. Special icing regions shown in Figures M-4, M-5, and M-6.
4. Areas where experience indicates unusual icing conditions exist.

In lieu of using the mapped values, it shall be permitted to determine the ice thickness, the concurrent wind speed, and the concurrent temperature for transmission line facilities from local meteorological data based on a 100-year MRI provided that:

1. The quality of the data for wind and type and amount of precipitation has been taken into account.
2. A robust ice accretion algorithm, with accreted ice assumed to remain on wires until the air temperature is above freezing, has been used to estimate ice thicknesses and concurrent wind speeds from these data.
3. Extreme-value statistical analysis procedures based on an extreme value distribution with at least three parameters have been employed in analyzing the ice thickness and concurrent wind speed data.
4. The length of record and sampling error has been taken into account.

M.10.2 Nominal Ice Thickness

Nominal ice thickness shall be determined using Figures M-2 through M-7. These figures show the equivalent uniform radial thicknesses of ice t due to freezing rain at a height of 33 ft (10 m) over the contiguous 48 states and Alaska for a 100-year MRI. Also shown are concurrent 3-second gust wind speeds. Thicknesses for Hawaii, and for ice accretions due to other sources in all regions, shall be obtained from local meteorological studies.

M.10.3 Loads Attributable to Freezing Rain with Concurrent Wind

M.10.3.1 Design Ice Thickness for Freezing Rain The design ice thickness t_z shall be calculated from Equations (M-14), (M-15), (M-16), or (M-17), as appropriate.

$$t_z = t_{100}(z/33)0.10 \text{ for } 0 \text{ ft} < z < 900 \text{ ft} \quad \text{(M-14)}$$

or

$$t_z = t_{\text{MRI}}(z/33)0.10 \text{ for } 0 \text{ ft} < z < 900 \text{ ft} \quad \text{(M-15)}$$

In SI:

$$t_z = t_{100}(z/10)0.10 \text{ for } 0 \text{ m} < z < 275 \text{ m} \quad \text{(M-16)}$$

or

$$t_z = t_{\text{MRI}}(z/10)0.10 \text{ for } 0 \text{ m} < z < 275 \text{ m} \quad \text{(M-17)}$$

where:

t_{100} = Nominal ice thickness due to freezing rain at a height of 33 ft (10 m) for 100-year MRI, from Figures M-2 through M-7 [in. (mm)],

t_{MRI} = Nominal ice thickness due to freezing rain at a height of 33 ft (10 m) at a selected MRI [inches (mm)],

t_z = Design ice thickness [in. (mm)], and

z = Height above ground [ft (m)]. For heights above ground greater than 900 ft (275 m), use z = 900 ft (275 m).

M.10.3.2 Ice Weight on Wires

The ice weight, W_i, shall be determined using Equation (M-18) considering a uniform thickness of glaze ice on the full length of all wires.

$$W_i = Q_i (d + t_z)\, t_z \qquad \text{(M-18)}$$

where:

W_i = Weight of glaze ice [lb/ft (N/m)],
Q_i = Constant to convert ice thickness to weight, 1.24 in customary units (0.0282 in metric units),
d = Diameter of bare wire [inches (mm)], and
t_z = Design ice thickness [inches (mm)].

The ice density used in Q_i shall not be less than 56 pcf (900 kg/m^3).

M.10.3.3 Ice Weight on Structures and Non-Structural Attachments Ice weight on transmission structures and non-structural attachments need not be considered for design of transmission structures. The weight of ice accretion due to freezing rain on non-structural attachments on wires such as aerial marker balls, bird diverters, and anti-galloping devices shall be considered in the calculation of the design wire tensions.

M.10.3.4 Wind on Structures and Non-Structural Attachments The wind pressure shall be determined using the concurrent wind speeds in Figures M-2 to M-7 and the procedures of Section M.9.1. The additional projected area due to ice accretion may be neglected when calculating the wind force on transmission structures and non-structural attachments.

M.10.3.5 Wind on Ice-Covered Wires Wind pressures applied to ice-covered wires shall be determined using the concurrent wind speeds in Figures M-2 to M-7 and the procedures of Section M.9.1.

M.10.3.6 Design Temperatures for Freezing Rain The design temperatures concurrent with the design ice and wind-on-ice loads due to freezing rain shall be the temperature for the site shown in Figures M-8 and M-9. Use either of these values or 32 °F (0 °C), whichever results in the maximum load effect. A temperature of 32 °F (0 °C) shall be used in Hawaii. These temperatures are applicable for ice thicknesses for all mean recurrence intervals.

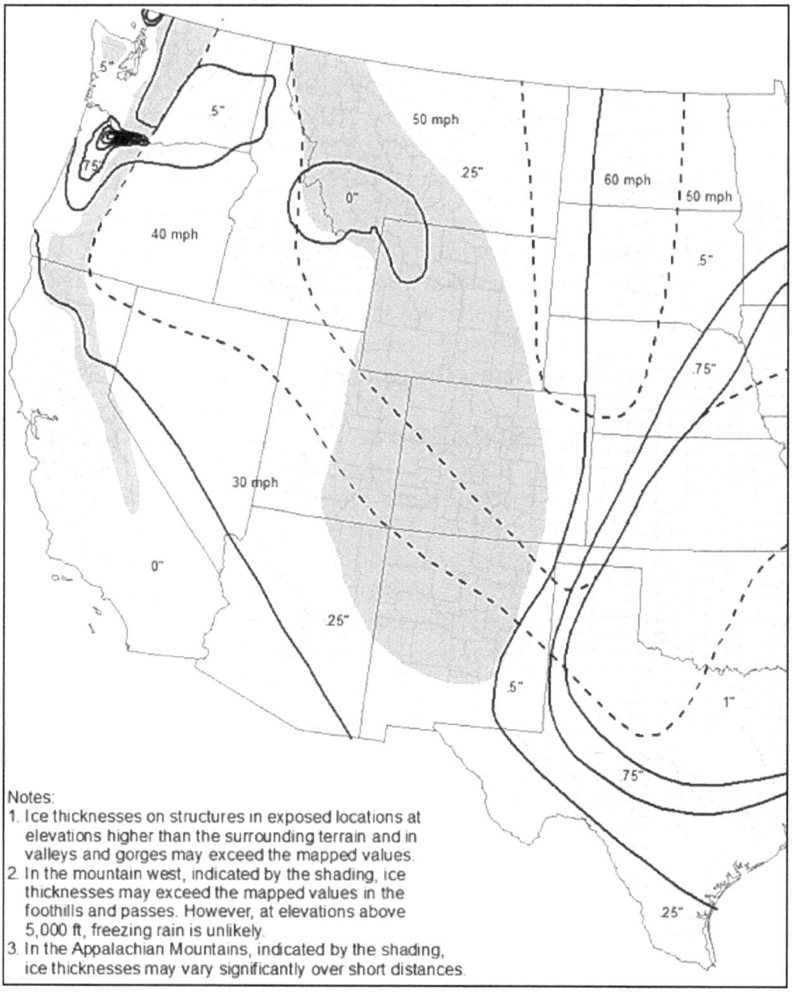

Figure M-2. 100-year MRI radial ice thickness (in.) from freezing rain with concurrent gust wind speeds (mph) at 33 ft (10 m) aboveground in Exposure C: Western United States.

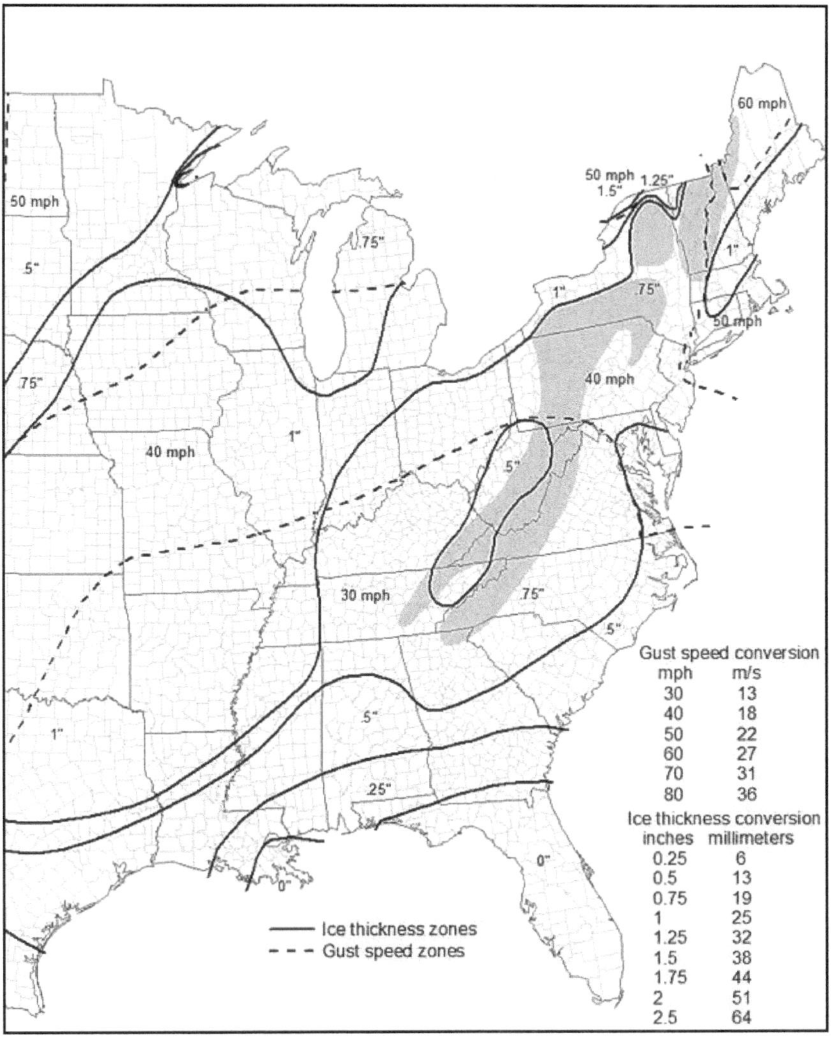

Figure M-3. 100-year MRI radial ice thickness (in.) from freezing rain with concurrent gust wind speeds (mph) at 33 ft (10 m) aboveground in Exposure C: Eastern United States.

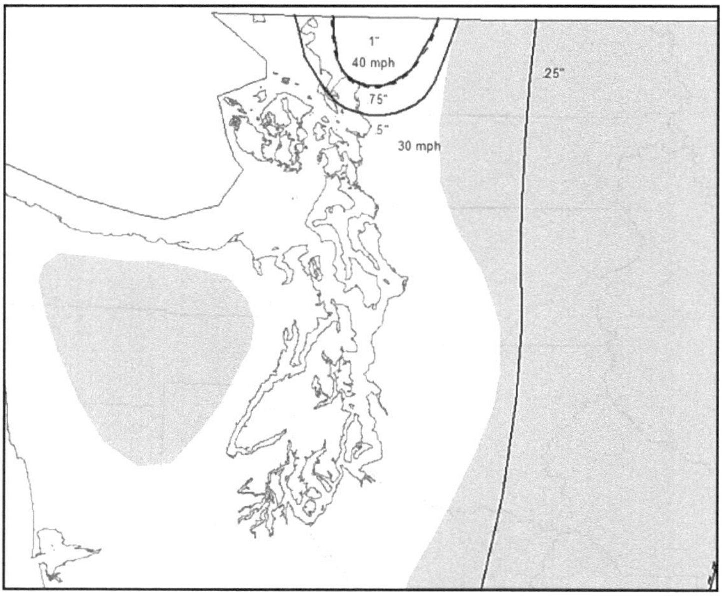

Figure M-4. 100-year MRI radial ice thickness (in.) from freezing rain with concurrent gust wind speeds (mph) at 33 ft (10 m) aboveground in Exposure C: Puget Sound detail.

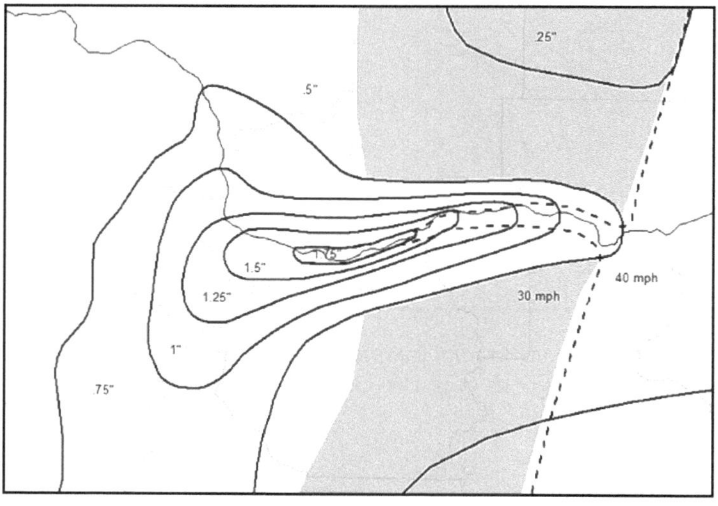

Figure M-5. 100-year MRI radial ice thickness (in.) from freezing rain with concurrent gust wind speeds (mph) at 33 ft (10 m) aboveground in Exposure C: Columbia River Gorge detail.

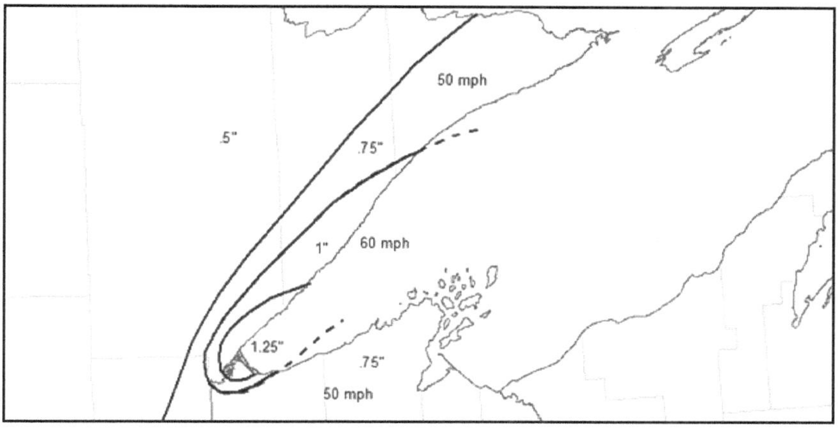

Figure M-6. 100-year MRI radial ice thickness (in.) from freezing rain with concurrent gust wind speeds (mph) at 33 ft (10 m) aboveground in Exposure C: Lake Superior detail.

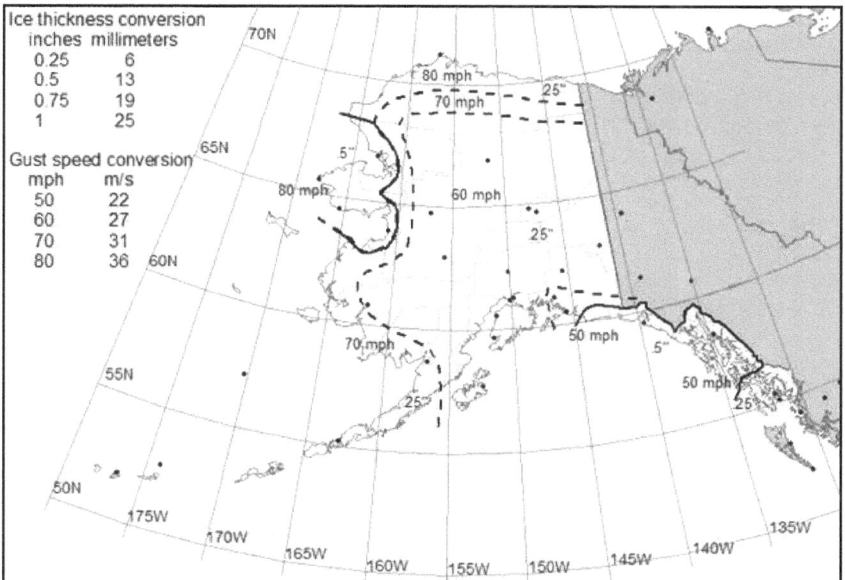

Figure M-7. 100-year MRI radial ice thickness (in.) from freezing rain with concurrent gust wind speeds (mph) at 33 ft (10 m) aboveground in Exposure C: Alaska.

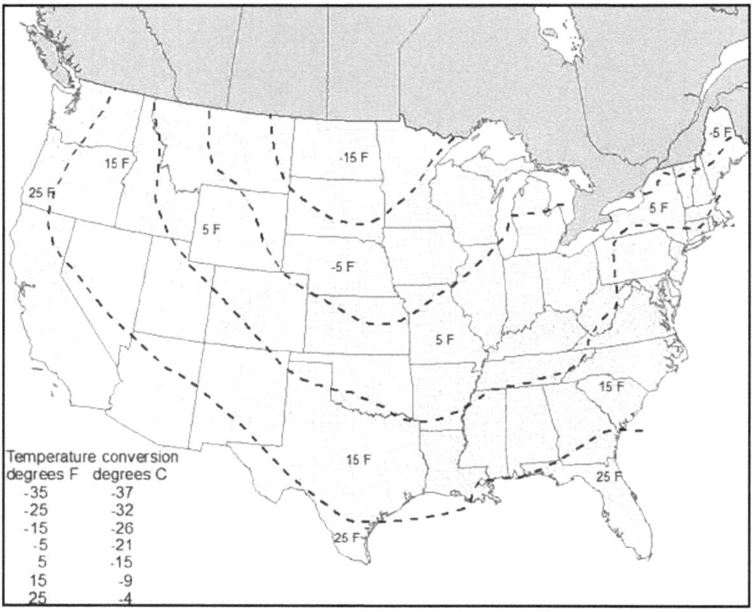

Figure M-8. Temperatures concurrent with ice thickness attributable to freezing rain (contiguous 48 states).

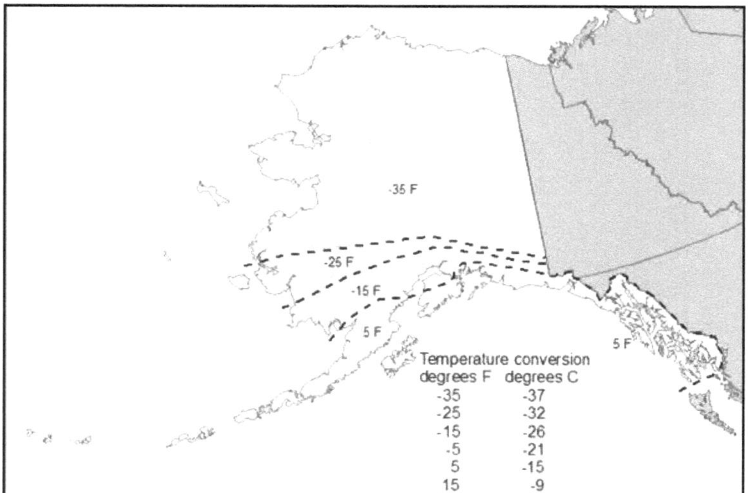

Figure M-9. Temperatures concurrent with ice thickness attributable to freezing rain: Alaska.

M.10.4 Unbalanced Ice Loading

The effects of unbalanced ice loading on the wires in adjacent spans shall be considered.

M.11 LEGISLATED LOADS

Transmission line facilities shall be designed to resist the loading specified by all applicable federal, state, and municipal codes, and legislative or administrative acts.

M.12 CONSTRUCTION AND MAINTENANCE LOADS

M.12.1 General

Loading from construction and maintenance activities shall be considered in the design of transmission line facilities. NOTE: For additional information on installation procedures, refer to the latest revisions of the following documents:

- IEEE Standard 524, *IEEE Guide to the Installation of Overhead Transmission Line Conductors;*
- IEEE Standard 951, *IEEE Guide to the Assembly and Erection of Metal Transmission Structures;*
- IEEE Standard 1025, *IEEE Guide to the Assembly and Erection of Concrete Pole Structures*; and
- Transmission Owner's construction specifications.

M.12.2 Climbing Loads

Loads acting on the structures resulting from workers installing or maintaining transmission facilities shall be considered in the design of the structure. Climbing anchorage attachment points and loads shall be considered for worker safety.

Minimum loading shall meet or exceed all Occupational Safety & Health Administration (OSHA) and other governmental requirements as applicable, and the latest revisions of IEEE 1307 the "IEEE Standards for Fall Protection for Utility Work" IEEE Standard 1307, and ANSI C2, *National Electrical Safety Code.*

M.13. OTHER HIGH-CONSEQUENCE EVENTS

Events with the potential for devastating damage to transmission line facilities resulting in extended outages shall be considered. Examples of these events include tornadoes, floods, landslides, avalanches, and sabotage.

REFERENCES

AASHTO. 1975. *Standard specifications for structural supports for highway signs, luminaires, and traffic signals*. Washington, DC: AASHTO.

Abbey, R. E., Jr. 1976. "Risk probabilities associated with tornado wind-speeds." In *Proc., Symp. on Tornadoes: Assessment of Knowledge and Implications for Man*, Lubbock, Texas: Texas Tech Univ., 177–236.

Abild, J., E. Y. Andersen, and L. Rosbjerg. 1992. "The climate of extreme winds at the Great Belt, Denmark." *J. Wind Eng. Ind. Aerodyn.* 41 (1–3): 521–532. https://doi.org/10.1016/0167-6105(92)90458-M.

Aboshosha, H., and A. El Damatty. 2013. "Span reduction factor of transmission line conductors under downburst winds." In *Proc., 8th Asia-Pacific Conf. on Wind Engineering*, Chennai, India. https://doi.org/10.3850/978-981-07-8012-8_P11.

ACI (American Concrete Institute). 2014. *Building code requirements for structural concrete and commentary*. ACI 318-14. Farmington Hills, MI: ACI.

AISC. 2010. *Specification for structural steel buildings*. AISC 360-10. Chicago: AISC.

ALCOA (Aluminum Company of America). 1961. *Graphic method for sag tension calculations for ACSR and other conductors*. Pittsburgh: ALCOA.

Anjo, K., S. Yamasaki, Y. Matsubayashi, Y. Nakayama, A. Otwsuki, and T. Fujimura. 1974. "An experimental study of bundle conductor galloping on the Kasatori-Yama test line for bulk power transmission." In *Proc., CIGRÉ 25th Session*, Vol. 1 (22–04), Paris.

ANSI (American National Standards Institute). 2002. *Specifications and dimensions for wood poles*. ANSI O5.1. New York: ANSI.

Armitt, J., M. Cojan, C. Manuzio, and P. Nicolini. 1975. "Calculation of wind loadings on components of overhead lines." In *Proc., Inst. Electr. Eng.* 122 (11): 1247–1252. https://doi.org/10.1049/piee.1975.0306.

Arya, S. P. S., M. E. Capuano, and L. C. Fagen. 1987. "Some fluid modeling studies of flow and dispersion over two dimensional hills." *Atmos. Environ.* 21 (4): 753–764. https://doi.org/10.1016/0004-6981(87)90071–0.

ASCE. 1961. "Wind forces on structures." *Trans. ASCE* 126 (2): 1124–1198.

ASCE. 1982. "Loadings for electrical transmission structures." *J. Struct. Div.*108 (ST5): 1088–1105.

ASCE. 1990a. *Design of steel transmission pole structures*. ASCE Manual of Practice No. 52. New York: ASCE.

ASCE. 1990b. *Minimum design loads for buildings and other structures*. ASCE 7-88. New York: ASCE.

ASCE. 1991. *Guidelines for electrical transmission line structural loading*. ASCE Manual of Practice No. 74, 2nd ed. New York: ASCE.

ASCE. 2000. *Design of latticed steel transmission structures*. ASCE 10-15. Reston, VA: ASCE.

ASCE. 2002. *Dynamic response of lattice towers and guyed masts*. Edited by M. K. S. Madugula. Reston, VA: ASCE.

ASCE. 2005. *Minimum design loads for buildings and other structures*. ASCE 7-05. Reston, VA: ASCE.

ASCE. 2006. *Reliability-based design of utility pole structures*. ASCE Manual of Practice No. 111. Reston, VA: ASCE.

ASCE. 2010a. *Guidelines for electrical transmission line structural loading*. ASCE Manual of Practice No. 74, 3rd ed. Reston, VA: ASCE.

ASCE. 2010b. *Minimum design loads for buildings and other structures*. ASCE 7-10. Reston, VA: ASCE.

ASCE. 2011. *Design of steel transmission pole structures*. ASCE 48-11. Reston, VA: ASCE.

ASCE. 2017. *Minimum design loads and associated criteria for buildings and other structures*. ASCE 7-16. Reston, VA.

Baenziger, M. A., S. Gupta, T. J. Wipf, F. Fanous, Y. H. Hahm, and H. B. White. 1993a. *Structural failure analysis of 345 kV transmission line, and discussion by B. White*. SM 441-6 PWRD. New York: IEEE.

Baenziger, M. A., W. D. James, B. Wouters, L. Li, and H. B. White. 1993b. *Dynamic loads on transmission line structures due to galloping conductors, and discussion by B. White*. WM 078-6 PWRD. New York: IEEE.

Baumeister, T., E. A. Avallone, T. Baumeister III, eds. (1978). *Standard handbook for mechanical engineers*, 8th ed. New York: McGraw-Hill, 9–167.

Bayar, D. C. 1986. "Drag coefficients of latticed towers." *J. Struct. Eng.* 112 (2): 417–430. https://doi.org/10.1061/(ASCE)0733-9445(1986)112:2(417).

BEAIRA (British Electrical and Allied Industries Research Association). 1935. "Wind pressure on latticed towers—Tests on models." *J. Inst. Electr. Eng.* 77 (464): 189–196. https://doi.org/10.1049/jiee-1.1935.0139.

Behncke, R. H., and T. C. E. Ho. 2009. "Review of span and gust factors for transmission line design." In *Proc., Electrical Transmission and Substation Structures 2009*, Fort Worth, TX. https://doi.org/10.1061/41077(363)18.

Bennett, I. 1959. *Glaze, its meteorology and climatology, geographical distribution and economic effects*. Technical Rep. EP-105. Natick, MA: US Army, Quartermaster Research and Engineering Center, Environmental Protection Research Division.

Bernstein, B. C., and B. G. Brown. 1997. "A climatology of supercooled large drop conditions based upon surface observations and pilot reports of icing." In *Proc., 7th Conf. on Aviation, Range and Aerospace Meteorology*, Long Beach, California.

Birjulin, A. P., V. V. Burgsdorf, and B. J. Makhlin. 1960. *Wind loads on overhead lines*. Paper 225. Paris: International Council on Large Electrical Systems (CIGRÉ).

Bocchieri, J. R. 1980. "The objective use of upper air soundings to specify precipitation type." *Mon. Weather Rev.* 108: 596–603. https://doi.org/10.1175/1520-0493(1980)108<0596:TOUOUA>2.0.CO;2.

Boddy, D. M., and J. Rice. 2009. "Impact of alternative galloping criteria on transmission line design." In *Proc., ASCE Electrical Transmission and Substation Structures 2009*, Fort Worth, Texas, 499–510.

Brekke, G. N. 1959. *Wind pressures in various areas of the United States*. Catalog No. C13.29. Washington, DC: US Weather Bureau, National Bureau of Standards.

Britter, R. E., J. C. R. Hunt, and K. J. Richards. 1981. "Air flow over a two-dimensional hill: Studies of velocity speed-up, roughness effects and turbulence." *Q. J. R. Meteorolog. Soc.* 107 (451): 91–110. https://doi.org/10.1002/qj.49710745106.

Brokenshire, R. E. 1979. *Experimental study of the loads imposed on welded steel support structures by galloping 345-kV bundled conductors*. Paper A 79 551-3. New York: IEEE.

Bryant, P, 2012. Limiting the Effects of Longitudinal Loads on Small Angle Lattice Transmission Towers, Philip Bryant, CenterPoint Energy, 2012.

BSI (British Standards Institution). 1972. *Basic data for the design of buildings—Chapter V: Loading—Part 2: Wind Loads*. BSI CP3. London: BSI.

Carper, K. L., ed. (1986). Forensic engineering: learning from failures. New York: ASCE.

Castanheta, M. N. 1970. *Dynamic behavior of overhead power lines subject to the action of the wind*. Paper No. 22-08. Paris: International Council on Large Electrical Systems (CIGRÉ).

CEA (Canadian Electrical Association). 1998. *Validation of ice accretion models for freezing precipitation using field data*. 331 T 992 (A-D). Montreal: CEA.

CENELEC (European Committee for Electrotechnical Standardization). 2001. *Overhead electrical lines exceeding AC 1 kV - Part 2-20: National Normative Aspects (NNA) for ESTONIA*. EN 50341. Brussels: CENELEC.

CENELEC. 2012. *Overhead electrical lines exceeding AC 1 kV. General requirements. Common specifications*. EN 50341. Brussels: CENELEC.

Chaîné, P. M., and G. Castonguay. 1974. *New approach to radial ice thickness concept applied to bundle-like conductors*. Industrial Meteorology—Study IV. Toronto: Environment Canada.

CIGRÉ (International Council on Large Electrical Systems). 2005. *Overhead conductor safe design tension with respect to Aeolian vibrations.* Task Force B2.11.04. Paris: CIGRÉ.

CIGRÉ. 2007. *State of the art conductor galloping.* Task Force B2.11.06. Paris: CIGRÉ.

CIGRÉ. 2009. *Overhead line design guidelines for mitigation of severe wind storm damage.* Scientific Committee B2 on Overhead Lines, B2.06.09. Paris: CIGRÉ.

Clough, P. W., and P. Penzien. 1975. *Dynamics of structures.* New York: McGraw-Hill.

Cluts, S., and A. Angelos. 1977. "Unbalanced forces on tangent transmission structures." In *Proc., IEEE Winter Power Meeting, New York,* Paper No. A77-220-7. New York: IEEE.

Colbeck, S. C., and S. F. Ackley. 1982. "Mechanisms for ice bonding in wet snow accretions on power lines." In *Proc., 1st Int. Workshop on Atmospheric Icing of Structures,* Hanover, New Hampshire, 25–30.

Comellini, E., and C. Manuzio. 1968. *Rational determination of design loadings for overhead line towers.* Paper 23-08. Paris: International Council on Large Electrical Systems (CIGRÉ).

Coppin, P. A., E. F. Bradley, and J. J. Finnigan. 1994. "Measurements of flow over an elongated ridge and its thermal stability dependence: The mean field." *Boundary Layer Meteorol.* 69 (1994): 173–199. https://doi.org/10.1007/BF00713302.

Davenport, A. G. 1960. *Wind loads on structures.* Technical Paper No. 88 of the Division of Building Research. Ottawa: National Research Council of Canada.

Davenport, A. G. 1962. "The response of slender line-like structures to a gusty wind." In *Proc., Inst. Civ. Eng.* 23 (3): 389–408. https://doi.org/10.1680/iicep.1962.10876.

Davenport, A. G. 1967. "Gust loading factors." *J. Struct. Div.* 93 (3): 11–34.

Davenport, A. G. 1977. "The prediction of the response of structures to gusty wind." In *Proc., Int. Research Seminar on the Safety of Structures under Dynamic Loading,* Trondheim, Norway, Vol. 1, 257–284.

Davenport, A. G. 1979. "Gust response factors for transmission line loading." In *Proc., 5th Int. Conf. on Wind Engineering (IAWE),* Fort Collins, CO, 899–909.

Davison et al. 1961. *Alcoa overhead engineering data no. 4.* Pittsburgh: Alcoa.

Den Hartog, J. P. 1932. "Transmission line vibration due to sleet." *Trans. AIEE* 51 (4): 1074–1086. https://doi.org/10.1109/T-AIEE.1932.5056223.

Durst, C. S. 1960. "Wind speeds over short periods of time." *Meteorol. Mag.* 89 (1056): 181–186.

ESDU (1993). "Strong winds in the atmospheric boundary layer. Part 2: Discrete gust speeds." ESDU Data Item No. 83045

Elawady, A., and A. A. El Damatty. 2018. "Critical load cases for lattice transmission line structures subjected to downbursts: Economic implications for design of transmission lines." *Eng. Struct.* 159: 213–226.

El Damatty, A. A., A. Hamada, and A. Elawady. 2013. "Development of critical load cases simulating the effect of downbursts and tornadoes on transmission line structures." In *Proc., 8th Asia-Pacific Conf. on Wind Engineering*, Chennai, India. https://doi.org/10.3850/978-981-07-8012-8_Key-01.

El Damatty, A. A., M. Hamada, and A. Hamada. 2015. "Simplified load cases for lattice transmission line structures under tornadoes." In *Proc., 14th Int. Conf. on Wind Engineering*, Porto Alegre, Brazil.

Engleman, W. C., and D. Marihugh. 1970. "Forces on conductors at high wind velocities." *Transmission and Distribution*, October.

EPRI (Electric Power Research Institute). 1978. *Longitudinal unbalanced loads on transmission line structures*. Rep. EL-643. Palo Alto, CA: EPRI.

EPRI. 1979. *Transmission line reference book*. Palo Alto, CA: EPRI.

EPRI. (1983). "Longitudinal unbalanced loads on transmission structures.", Computer Programs Documentation, BROD12 and BROFLX, Project 561-1 Report EL-2943-CC7.

ESDU (Engineering Science Data Unit). 1982. *Strong winds in the atmospheric boundary layer. Part 1: Mean hourly wind speed*. ESDU 82026. London: ESDU.

ESDU. 1983. *Strong winds in the atmospheric boundary layer. Part 2: Discrete gust speeds*. ESDU 83045. London: ESDU.

Farr, H. H. 1980. *Transmission line design manual*. Denver: US Dept. of the Interior, Water, and Power Resources Service.

Felin, B. 1988. "Freezing rain in Quebec: Field observations compared to model estimations." In *Proc., 4th Int. Workshop on Atmospheric Icing of Structures*, Paris, 119–123.

Finnigan, J. J., M. R. Raupach, E. F. Bradley, and G. K. Aldis. 1990. "A wind tunnel study of turbulent flow over a two-dimensional ridge." *Boundary Layer Meteorol.* 50 (1990): 277–317. https://doi.org/10.1007/BF00120527.

Finstad, K., R. Hopkinson, and K. Jones. 2009. *Wet snow modeling for Alberta—2009*. Rep. for Alberta Electric System Operator (AESO). Ottawa: VP Enterprises.

Finstad, K., and E. Lozowski. 1988. "A computational investigation of water droplet trajectories." *J. Atmos. Oceanic Technol.* 5: 160–170. https://doi.org/10.1175/1520-0426(1988)005<0160:ACIOWD>2.0.CO;2.

Frandsen, A. G., and P. H. Juul. 1976. *Cascade of tower collapses: Design criteria*. Paper 22-10. Paris: International Council on Large Electrical Systems (CIGRÉ).

Fujita, T. T. 1985. *The downburst: Microburst and microburst*. SMRP Research Paper 210. Chicago: Univ. of Chicago.

Fujita, T. T., and A. D. Pearson. 1973. "Results of FPP classification of 1971 and 1972 tornadoes." In *Proc., 8th Conf. on Severe Local Storms*, Denver, 142–145.

Gibbon, R. R. 1984. *Damage to overhead lines caused by conductor galloping.* Art. ELT_094_4. Paris: International Council on Large Electrical Systems (CIGRÉ).

Gland, H., and P. Admirat. 1986. "Meteorological conditions for wet snow occurrence in France—Calculated and measured results in a recent case study on 5 March 1985." In *Proc., 3rd Int. Workshop on Atmospheric Icing of Structures*, Vancouver, 91–96.

Gong, W., and A. Ibbetson. 1989. "A wind tunnel study of turbulent flow over model hills." *Boundary-Layer Meteorol.* 49 (1989): 113–148. https://doi.org/10.1007/BF00116408.

Goodwin, E. J., J. D. Mozer, A. M. DiGioia Jr., and B. A. Power. 1983. "Predicting ice and snow loads for transmission line design." In *Proc, 3rd Int. Workshop on Atmospheric Icing of Structures*, Vancouver, 267–275.

Gouze, S. C., and M. C. Richmond. 1982a. *Meteorological evaluation of the proposed Alaska transmission line routes.* Altadena, CA: Meteorology Research, Inc.

Gouze, S. C., and M. C. Richmond. 1982b. *Meteorological evaluation of the proposed Palmer to Glennallen transmission line route.* Altadena, CA: Meteorology Research, Inc.

Griffing, K. L., and Leavengood, D. C. (1973). "Transmission line failures, part I. Meteorological phenomena— severe winds and icing." In *Proc., IEEE Winter Power Mtg.*, Paper No. 73 CH0816-9-PWR, 5–14. New York.

Hamada, A., and A. A. El Damatty. 2011. "Behaviour of guyed transmission line structures under tornado wind loading." *Comput. Struct.* 89 (11–12): 986–1003. https://doi.org/10.1016/j.compstruc.2011.01.015.

Hamada, A., and A. A. El Damatty. 2015. "Failure analysis of guyed transmission lines during F2 tornado event." *Eng. Struct.* 85 (February 15): 11–25. https://doi.org/10.1016/j.engstruct.2014.11.045.

Hamada, A., A. A. El Damatty, H. Hangan, and A. Y. Shehata. 2010. "Finite element modelling of transmission line structures under tornado wind loading." *Wind Struct.* 13 (5): 451–469.

Hangan, H., and J.-D. Kim. 2008. "Swirl ratio effects on tornado vortices in relation to the Fujita scale." *Wind Struct.* 11 (4): 291–302.

Hangan, H., D. Roberts, Z. Xu, and J. Kim. 2003. "Downburst simulation. Experimental and numerical challenges." In *Proc., 11th Int. Conf. on Wind Engineering*, Lubbock, Texas, June 1–5.

Havard, D. G., and J. C. Pohlman. 1980. *Control of galloping conductors by detuning.* Paper 22-05. Paris: International Council on Large Electrical Systems (CIGRÉ).

Havard, D. G., A. S. Paulson, and J. C. Pohlman. 1982. *The economic benefits of controls for conductor galloping*. Paper 22-02. Paris: International Council on Large Electrical Systems (CIGRÉ).

Hjelmfelt, M. R. 1988. "Structure and life cycle of microburst outflows observed in Colorado." *J. App. Meteorol.* 27 (8): 900–927. https://doi.org/10.1175/1520-0450(1988)027<0900:SALCOM>2.0.CO;2.

Hoerner, S. F. 1958. *Fluid-dynamic drag*. Vancouver, WA: Hoerner Fluid Dynamics.

Holmes, J. D. 2001. *Wind loading of structures*. New York: Spon.

Holmes, J. D., H. M. Hangan, J. L. Schroeder, C. W. Letchford, and K. D. Orwig. 2008. "A forensic study of the Lubbock-Reese downdraft of 2002." *Wind Struct.* 11 (2): 137–152. https://doi.org/10.12989/was.2008.11.2.137.

Hoskings, J. R. M., and J. R. Wallis. 1987. "Parameter and quantile estimation for the generalized Pareto distribution." *Technometrics* 29 (3): 339–349. https://doi.org/10.2307/1269343.

Houghton, E. L., and N. B. Carruthers. 1976. *Wind forces on building and structures.* New York: Wiley.

IEC (International Electrotechnical Commission). 2003a. *Design criteria of overhead transmission lines.* IEC 60826, Ed. 3.0. Geneva: IEC.

IEC. 2003b. *Loading and strength of overhead transmission lines.* IEC 60826, 3rd ed. Geneva: IEC.

IEC. 2017. *Design criteria of overhead transmission lines.* IEC 60826, Ed. 4.0. Geneva: IEC.

IEEE. 1999. "Limitations of the ruling span method for overhead line conductors at high operating temperatures." *IEEE Trans. Power Deliv.* 14 (2): 549–560. https://doi.org/10.1109/61.754102.

IEEE. 2003. *IEEE guide to the installation of overhead transmission line conductors.* IEEE 524. New York: IEEE.

IEEE. 2004. *IEEE standard for fall protection for utility work.* IEEE 1307. New York: IEEE.

IEEE. 2012. *National electrical safety code.* ANSI C2. New York: IEEE.

IEEE. 2017. *National electrical safety code.* ANSI C2-2017. New York: IEEE.

ISO (International Standards Organization). 1999. *Atmospheric icing of structures.* ISO 12494. Geneva: ISO.

James, W. D. 1976. "Effects of Reynolds number and corner radius on two-dimensional flows around octagonal, dodecagonal, and hexdecagonal cylinders." Ph.D. diss., College of Engineering, Iowa State Univ.

Janney, J. R. (1979). Guide to investigation of structural failures., New York: ASCE.

Jones, K. 1996. "A simple model for freezing rain loads." In *Proc., 7th Int. Workshop on Atmospheric Icing of Structures,* Chicoutimi, Quebec, 412–416.

Jones, K. F. 1998. "A simple model for freezing rain ice loads." *Atmos. Res.* 46 (1–2): 87–97. https://doi.org/10.1016/S0169-8095(97)00053-7.

Jung, S., and F. J. Masters. 2013. "Characterization of open and suburban boundary layer wind turbulence in 2008 Hurricane Ike." *Wind Struct.* 17 (2): 135–162. https://doi.org/10.12989/was.2013.17.2.135.

Kempner, L., Jr. 1997. "Longitudinal impact loading on electrical transmission line towers—A scale model study." Ph.D. diss., Dept. of Civil and Environmental Engineering, Portland State Univ.

Keselman, L. M., and Y. Motlis. 1996. "Enhanced analytical design method for overhead line conductors in non-level spans." In *Proc., IEEE/PES Transmission and Distribution Conf.*, Los Angeles, 359–366.

Keselman, L. M., and Y. Motlis. 1998. "Application of the ruling span concept for overhead lines in mountainous terrain." *IEEE Trans. Power Deliv.* 13 (4): 1385–1390. https://doi.org/10.1109/61.714512.

Kuroiwa, D. 1962. *A study of ice sintering*. Research Rep. 86. Hanover, NH: US Army Cold Regions Research and Engineering Laboratory.

Kuroiwa, D. 1965. *Icing and snow accretion on electric wires*. Research Report 123. Hanover, NH: US Army Cold Regions Research and Engineering Laboratory.

Langmuir, I., and K. Blodgett. 1946. "Mathematical investigation of water droplet trajectories." US Army Air Forces Tech. Rep. 5418. Washington, DC: Office of the Publication Board, Dept. of Commerce.

MacDonald, A. 1975. *Wind loading on buildings*. New York: Wiley.

Makkonen, L. 1996. "Modeling power line icing in freezing precipitation." In *Proc., 7th Int. Workshop on Atmospheric Icing of Structures*, Chicoutimi, Quebec, 195–200.

Mara, T. G. 2014. "Influence of solid area distribution on the drag of a two-dimensional lattice frame." *J. Eng. Mech.* 140 (3): 644–649. https://doi.org/10.1061/(ASCE)EM.1943-7889.0000681.

Mara, T. G. 2015. "Updated gust response factors for transmission line loading." In *Proc., Electrical Transmission and Substation Structures Conf. 2015*, Branson, Missouri. https://doi.org/10.1061/9780784479414.037.

Mara, T. G., J. K. Galsworthy, and E. Savory. 2010. "Assessment of vertical wind loads on lattice framework with application to thunderstorm winds." *Wind Struct.* 13 (5): 413–431. https://doi.org/10.12989/was.2010.13.5.413.

Mathur, R. K., A. H. Shah, P. G. S. Trainor, and N. Popplewell. 1986. *Dynamics of guyed transmission tower system*. Paper 86 SM 414-7. New York: IEEE.

McComber, P., R. Martin, G. Morin, and L. V. Van. 1982. "Estimation of combined ice and wind load on overhead transmission lines." In *Proc., 1st Int. Workshop on Atmospheric Icing of Structures*, Hanover, New Hampshire, 143–153.

McCormick, T., and J. C. Pohlman. 1993. "Study of compact 220 kV line system indicates need for micro-scale meteorological information." In *Proc., 6th Int. Workshop on Atmospheric Icing of Structures*, Budapest.

McDonald, J. R. 1983. *A methodology for tornado hazard probability assessment*. NUREG/CR-3058. Washington, DC: Nuclear Regulatory Commission.

Mehta, K. C., and T. Lou. 1983. *Force coefficients for transmission towers*. Electric Power Research Institute (EPRI) Rep., Technical Agreement TPS 82–623. Washington, DC: EPRI.

Mehta, K. C., J. E. Minor, and J. R. McDonald. 1976. "Windspeeds analyses of April 3–4, 1974 tornadoes." *J. Struct. Div.* 102 (9): 1709–1724.

Meteorological Research Institute. 1977. *Ontario Hydro wind and ice loading model*. MRI 77 FR-1496. Toronto: Ontario Hydro.

Miller, C. A. 2011. "Revisiting the Durst gust factor curve." *Can. J. Civ. Eng.* 38 (9): 998–1001.

Miller, C. A., and A. G. Davenport. 1998. "Guidelines for the calculation of wind speed-ups in complex terrain." *J. Wind Eng. Ind. Aerodyn.* 74–76 (April 1): 189–197. https://doi.org/10.1016/S0167-6105(98)00016-6.

Minor, J. E., J. R. McDonald, and K. C. Mehta. 1977. *The tornado: An engineering-oriented perspective*. NOAA Technical Memorandum ERL NSSL-82. Norman, OK: National Severe Storms Laboratory.

Modi, V. J., and J. E. Slater. 1983. "Unsteady aerodynamics and vortex-induced aeroelastic instability of a structural angle section." *J. Wind Eng. Ind. Aerodyn.* 11 (1–3): 321–334.

Mozer, J. D., J. C. Pohlman, and J. F. Fleming. 1977. "Longitudinal load analysis of transmission line system." *IEEE Trans. Power Apparatus Syst.* 96 (5): 1657–1665. https://doi.org/10.1109/T-PAS.1977.32495.

Mozer, J. D., and R. J. West. 1983. "Analysis of 500 kV tower failures." In *Proc., Pennsylvania Electric Association*, Norristown, Pennsylvania.

National Bureau of Standards. 1920. *National electrical safety code*. Handbook No. 3. Washington, DC: US Dept. of Commerce.

Nebraska Public Power District. 1976. "The storm of March 29, 1976." Columbus, NE: Nebraska Public Power District.

Nigol, O., and Havard, D. G. 1978. *Control of torsionally induced conductorsby detuning*. Paper A 78 125-7. Piscataway, NJ: IEEE.

NOAA (National Oceanic and Atmospheric Administration). (1959–Present). Storm data. Accessed June 10, 2020. https://www.ncdc.noaa.gov/stormevents/.

NRCC (National Research Council of Canada). 2010. *National building code of Canada (NBCC)*. Ottawa: NRCC.

Orwig, K. D. and J. L. Schroeder. 2007. "Near-surface wind characteristics of severe thunderstorm outflows." *J. Wind Eng. Ind. Aerodyn.* 95 (7): 565–584. https://doi.org/10.1016/j.jweia.2006.12.002.

Ostendorp, M. 1997. *Longitudinal load and cascading failure risk assessment (CASE)*. Vol. I–IV of EPRI Rep. TR-107087. Palo Alto, CA: Electric Power Research Institute (EPRI).

Paz, M. 1980. *Structure dynamics—Theory and computations*. New York: Van Nostrand Reinhold.

Peabody, A. B. 1993. "Snow loads on transmission and distribution lines in Alaska." In *Proc., Sixth Int. Workshop on Atmospheric Icing of Structures*, Budapest.

Peterka, J. A. 1992. "Improved extreme wind prediction for the United States." *J. Wind Eng. Ind. Aerodyn.* 41 (1–3): 533–541. https://doi.org/10.1016/0167-6105(92)90459-N.

Peterka, J. A., K. Finstad, and A. K. Pandy. 1996. *Snow and wind loads for Tyee transmission line*. Fort Collins, CO: Cermak Peterka Petersen.

Peyrot, A. H. 1985. "Microcomputer-based nonlinear structural analysis of transmission line systems." *IEEE Trans. Power Apparatus Syst.* PAS-104 (11): 3236–3244. https://doi.org/10.1109/TPAS.1985.318837.

Peyrot, A. H. and H. J. Dagher. 1984. "Reliability-based design of transmission lines." *J. Struct. Eng.* 110 (11): 2758–2777. https://doi.org/10.1061/(ASCE)0733-9445(1984)110:11(2758).

Pintar, A. L., E. Simiu, F. T. Lombardo, and M. Levitan. 2015. *Maps of non-hurricane non-tornadic wind speeds with specified mean recurrence intervals for the contiguous United States using a two-dimensional Poisson process extreme value model and local regression*. NIST Special Publication 500-301. Gaithersburg, MD: NIST.

Pohlman, J. C., and D. Havard. 1979. *Field research on the galloping of iced conductors—A status report*. Paper A 79 551-3. New York: IEEE.

Pohlman, J. C., and C. B. Rawlins. 1979. "Field testing an anti-galloping concept on overhead lines." In *Proc., 7th IEEE/PES Transmission and Distribution Conference and Exposition*, Atlanta. https://doi.org/10.1109/TDC.1979.712640.

Rawlins, C. B. 1981. "Analysis of conductor galloping field observations single conductors." *IEEE Trans. Power Apparatus Syst.* PAS-100 (8): 3744–3751. https://doi.org/10.1109/TPAS.1981.317017.

REA (Rural Electrification Association). 1980. "Design manual for high voltage transmission lines," Bulletin 62-I. Washington, DC: US Dept. of Agriculture, Rural Electrification Administration.

Richards, D. J. W. 1965. "Aerodynamic properties of the Severn crossing conductor." In *Proc., Symp. 16*, Middlesex, United Kingdom, Paper 8.

Richardson, A. S. 1983. *The wind damper method of galloping control*. DOE/CE/15102T1. Springfield, VA: National Technical Service, US Dept. of Commerce.

Richardson, A. S. 1986. *Longitudinal dynamic loading of steel pole transmission lines*. Paper WM 189-5. New York: IEEE.

Richmond, M. C. 1985. *Meteorological evaluation of Bradley Lake hydroelectric project 115 kV transmission line route*. Torrance, CA: M. C. Richmond Meteorological Consulting.

Richmond, M. C. 1991. *Meteorological evaluation of Tyee Lake hydroelectric project transmission line route, Wrangell to Petersburg*. Torrance, CA: M.C. Richmond Meteorological Consulting.

Richmond, M. C. 1992. *Meteorological evaluation of Tyee Lake hydroelectric project transmission line route, Tyee power plant to Wrangell*. Torrance, CA: M.C. Richmond Meteorological Consulting.

Riley, M. J., L. Kempner Jr., W. H. Mueller III, and G. T. Gobo. 2002. "A comparison of seismic (dynamic) and static load cases for lattice electric transmission towers." In *Proc., ASCE Electrical Transmission in a New Age*, Omaha, Nebraska, 420–426. https://doi.org/10.1061/40642(253)39.

Robbins, C. C., and J. V. Cortinas Jr. 1996. "A climatology of freezing rain in the contiguous United States: Preliminary results." In *Preprints, 15th AMS Conf. on Weather Analysis and Forecasting*, American Meteorological Society, Norfolk, Virginia.

Sachs, P. 1972. *Wind forces in engineering*. New York: Pergamon.

Sachs, P. 1978. *Wind forces in engineering*, 2nd ed. New York: Pergamon.

Sakamoto, Y., K. Mizushima, and S. Kawanishi. 1990. "Dry snow type accretion on overhead wires: growing mechanism, meteorological conditions under which it occurs and effect on power lines." In *Proc., 5th Int. Workshop on Atmospheric Icing of Structures*, Tokyo, Paper 5-9.

Scruton, C. and C. W. Newberry. 1963. "On the estimation of wind loads on buildings and structural design." In *Proc., Inst. Civil Eng.* 25 (6): 97–126. https://doi.org/10.1680/iicep.1963.10660.

Shan, L., and L. Marr. 1996. *Ice storm data base and ice severity maps*. Palo Alto, CA: Electric Power Research Institute (EPRI).

Shehata, A. Y., and A. A. El Damatty. 2007. "Behaviour of guyed transmission line structures under downburst wind loading." *Wind Struct.* 10 (3): 249–268. https://doi.org/10.12989/was.2007.10.3.249.

SIA (Swiss Association of Engineers and Architects). 1956. *Standards for load assumptions, acceptance and inspection of structures*. Tech. Paper No. 160. Zurich: SIA.

Simiu, E., and R. H. Scanlan. 1996. *Wind effects on structures: Fundamentals and applications to design*, 3rd ed. New York: Wiley.

Slater, J. E., and V. J. Modi. 1971. "On the wind-induced vibration of structural angle sections." In *Proc., IAWE, Third Int. Conf. on Wind Effects on Building and Structures*, Tokyo.

Snyder, W. H., and R. E. Britter. 1987. "A wind tunnel study of the flow structure and dispersion from sources upwind of three-dimensional hills, atmospheric environment." *Atmos. Environ.* 21 (4): 735–751. https://doi.org/10.1016/0004-6981(87)90070-9.

Stallabrass, J. R. and P. F. Hearty. 1967. *The icing of cylinders in conditions of simulated freezing sea spray*. Mechanical Engineering Rep. MD-50, NRC No. 9782. Ottawa: National Research Council.

Tattelman, P., and I. Gringorten. 1973. *Estimated glaze ice and wind loads at the earth's surface for the contiguous United States*. Rep. AFCRL-TR-73-0646. Bedford, MA: US Air Force Cambridge Research Laboratories.

Taylor, P. A., P. J. Mason, and E. F. Bradley. 1987. "Boundary layer flow over low hills, a review." *Boundary Layer Meteorol.* 39 (1–2): 107–132. https://doi.org/10.1007/BF00121870.

Tecson, J. J., T. T. Fujita, and R. F. Abbey Jr. 1979. "Statistics of U. S. tornadoes based on the DAPPLE tornado tape." In *Proc., 11th Conf. on Severe Local Storms,* Kansas City, Missouri, 227–234.

Thayer, E. S. 1924. "Computing tensions in transmission lines." *Electr. World* 84 (2): 72–73.

Thomas, M. B. 1981. "Broken conductor loads on transmission line structures," Ph.D. diss., College of Civil and Environmental Engineering, Univ. of Wisconsin–Madison.

Thrasher, W. J. 1984. "Halt redundant member failure of 765-KV towers." *Trans. Distrib.,* May.

Trainor, P. G. S., N. Popplewell, A. H. Shah, and C. K. Wong. 1984. *Estimation of fundamental dynamic characteristics of transmission towers.* Paper 84 SM 525-2. New York: IEEE.

Twisdale, L. A. 1982. "Wind-loading underestimate in transmission line design." *Trans. Distrib.,* December, 40–46.

USFS (United States Forest Service). 1994. *Telecommunications handbook, R3 supplement 6609.14-94-2.* Forest Service Handbook, FSH 6609, 14, Washington, DC: USFS.

Vellozzi, J. W., and E. Cohen. 1968. "Gust response factors." *J. Struct. Div.* 94 (6): 1295–1314.

Vickery, P. J., and P. F. Skerlj. 2005. "Hurricane gust factors revisited." *J. Struct. Eng.* 131 (5): 825–832. https://doi.org/10.1061/(ASCE)0733-9445(2005)131:5(825).

Vickery, P. J., D. Wadhera, J. Galsworthy, J. A. Peterka, P. A. Irwin, and L. A. Griffis. 2010. "Ultimate wind load design gust wind speeds in the United States for use in ASCE-7." *J. Struct. Eng.* 136 (5): 613–625. https://doi.org/10.1061/(ASCE)ST.1943-541X.0000145.

Walmsley, J. L., P. A. Taylor, and T. Keith. 1986. "A simple model of neutrally stratified boundary-layer flow over complex terrain with surface roughness modulations (MS3DJH/3R)." *Boundary-Layer Meteorol.* 36 (1986): 157–186. https://doi.org/10.1007/BF00117466.

Wang, Q. J. 1991. "The POT model described by the generalized Pareto distribution with Poisson arrival rate." *J. Hydrol.* 129 (1–4): 263–280. https://doi.org/10.1016/0022-1694(91)90054-L.

Wardlaw, R. L. 1967. "Wind-excited vibrations of slender beams with angle cross-sections." In *Proc., Int. Research Seminar—Wind Effects on Buildings and Structures*, Ottawa.

Watson, L. T. 1955. *Drag of bare stranded cables.* Tech. Memorandum 114. Melbourne, AU: Aeronautical Research Laboratories.

Wen, Y. K., and S.-L. Chu. 1973. "Tornado risks and design wind speed." *J. Struct. Div.* 99 (12): 2409–2421.

Whapham, R. 1982. *Field research for control of galloping with air flow spoilers.* Cleveland: Preformed Line Products.

Whitbread, R. S. 1979. "The influence of shielding on the wind forces experienced by arrays of lattice frames." In *Proc., 5th Int. Conf. on Wind Engineering (IAWE)*, Fort Collins, Colorado, 405–420.

White, H. B. 1979. *Some destructive mechanisms activated by galloping conductors.* Report No. A79 106-6. New York: IEEE.

White, H. B. 1999. "Galloping of ice covered wires of a transmission line." In *Proc., ASCE Cold Regions Conf.*, Lincoln, New Hampshire, 790–798.

Winkelman, P. F. 1959. *Sag-tension computations and field measurements of Bonneville Power Administration.* AIEE Paper 59-900. New York: American Institute of Electrical Engineers.

Yip, T.-C. 1995. "Estimating icing amounts caused by freezing precipitation in Canada." *Atmos. Res.* 36 (3–4): 221–232. https://doi.org/10.1016/0169-8095(94)00037-E.

Yip, T. C., and P. Mitten. 1991. *Comparisons between different ice accretion models.* Ottawa: Canadian Electrical Association.

Young, W. R. 1978. "Freezing precipitation in the southeastern United States." M.S. diss., Dept. of Atmospheric Sciences, Texas A&M Univ.

ADDITIONAL RESOURCES

Bryant, P. J. 2012. "Limiting the effects of longitudinal loads on small angle lattice transmission towers." In *Proc., ASCE Electrical Transmission and Substation Structures 2012*, Columbus, Ohio, 205–216. https://doi.org/10.1061/9780784412657.018.

Elawady, A., and A. A. El Damatty. 2016. "Longitudinal force on transmission structures due to non-symmetric downburst conductor loads." *Eng. Struct.* 127 (Nov): 206–226. https://doi.org/10.1016/j.engstruct.2016.08.030.

Freimark, B., et al. 2012. "The effect of broken wire loads on EHV transmission structure design." In *Proc., ASCE Electrical Transmission and Substation Structures Conf. 2012*, Columbus, OH, 166–182. https://doi.org/10.1061/9780784412657.015.

Mara, T. G., and R. H. Behncke. 2015. "Examination of yawed wind loading on transmission towers." In *Proc., Electrical Transmission and Substation Structures Conf. 2015*, Branson, MO. https://doi.org/10.1061/9780784479414.042.

Peabody, A. B. 2004. "Applying shock damping to the problem of transmission line cascades." Ph.D. thesis, Dept. of Civil Engineering and Applied Mechanics, McGill Univ.

INDEX